Lateral thinking

Other books by Edward de Bono

The use of lateral thinking
The five-day course in thinking
The mechanism of mind

Lateral thinking:
creativity step by step
Edward de Bono

HARPER COLOPHON BOOKS
Harper & Row, Publishers
New York, Hagerstown, San Francisco, London

This book was originally published in a hardcover edition in 1970 by Harper & Row, Publishers.

First HARPER COLOPHON edition published 1973

ISBN: 0 −06 −090325 −2

83 84   20 19 18 17 16 15 14 13

Contents

Preface

This book is intended for use both at home and at school. At school the emphasis has traditionally always been on vertical thinking which is effective but incomplete. This selective type of thinking needs to be supplemented with the generative qualities of creative thinking. This is beginning to happen in some schools but even so creativity is usually treated as something desirable which is to be brought about by vague exhortation. There is no deliberate and practical procedure for bringing it about. This book is about lateral thinking which is the process of using information to bring about creativity and insight restructuring. Lateral thinking can be learned, practised and used. It is possible to acquire skill in it just as it is possible to acquire skill in mathematics.

The book should be of use to teachers who are looking for a practical way to handle this type of thinking which is becoming ever more important. The book provides formal opportunities to practise lateral thinking and also an explanation of the processes involved. The teacher may either use the book for his or her own interest or, better still, as a basis for classroom work.

Since the universal introduction of practical creativity into school education may take some time to come about, parents might not wish to wait for this. They might prefer to supplement the school situation with home instruction in lateral thinking.

It is emphasized that there is no antagonism between the two sorts of thinking. Both are necessary. Vertical thinking is immensely useful but one needs to enhance its usefulness by adding creativity and tempering its rigidity. Eventually this will be done at school but until that time it may be necessary to do it at home.

The book is not intended to be read through at one

sitting but worked through slowly – over months or even years. For that reason many of the principles are repeated at intervals throughout the book in order to hold the subject together and prevent it fragmenting into mere techniques. In using the book it is important to remember that practice is far more important than understanding of the process.

Introduction

Lateral thinking is closely related to insight, creativity
and humour. All four processes have the same basis.
But whereas insight, creativity and humour can only be
prayed for, lateral thinking is a more deliberate process.
It is as definite a way of using the mind as logical
thinking—but a very different way.

Culture is concerned with establishing ideas. Education
is concerned with communicating those established
ideas. Both are concerned with improving ideas by
bringing them up to date. The only available method
for changing ideas is conflict which works in two ways.
In the first way there is a head on confrontation between
opposing ideas. One or other of the ideas achieves a
practical dominance over the other idea which is
suppressed but not changed. In the second way there is
a conflict between new information and the old idea. As
a result of this conflict the old idea is supposed to be
changed. This is the method of science which is always
seeking to generate new information to upset the old
ideas and bring about new ones. It is more than the
method of science—it is the method of human
knowledge.

Education is based on the safe assumption that one only
has to go on collecting more and more information for it
to sort itself into useful ideas. We have developed tools
for handling the information: mathematics for
extending it, logical thinking for refining it.

The conflict method for changing ideas works well
where the information can be evaluated in some
objective manner. But the method does not work at all
when the new information can only be evaluated
through the old idea. Instead of being changed the old
idea is strengthened and made ever more rigid.

The most effective way of changing ideas is not from

outside by conflict but from within by the insight
rearrangement of available information. Insight is the
only effective way of changing ideas in a myth situation
– when information cannot be evaluated objectively.
Even when information can be evaluated objectively, as
in science, an insight rearrangement of information
leads to huge leaps forward. Education is not only
concerned with collecting information but also with the
best ways of using information that has been collected.

When ideas lead information rather than lag behind
progress is rapid. Yet we have developed no practical
tools for handling insight. We can only go on collecting
information and hope that at some stage it will come
about. Lateral thinking is an insight tool.

Insight, creativity and humour are so elusive because
the mind is so efficient. The mind functions to create
patterns out of its surroundings. Once the patterns are
formed it becomes possible to recognize them, to react
to them, to use them. As the patterns are used they
become ever more firmly established.

The pattern using system is a very efficient way of
handling information. Once established the patterns
form a sort of code. The advantage of a code system is
that instead of having to collect all the information one
collects just enough to identify the code pattern which
is then called forth even as library books on a particular
subject are called forth by a catalogue code number.

It is convenient to talk of the mind as if it were some
information handling machine – perhaps like a computer.
The mind is not a machine however, but a special
environment which allows information to organize
itself into patterns. This self-organizing,
self-maximizing, memory system is very good at
creating patterns and that is the effectiveness of mind.

But inseparable from the great usefulness of a patterning
system are certain limitations. In such a system it is easy
to combine patterns or to add to them but it is extremely
difficult to restructure them for the patterns control
attention. Insight and humour both involve the
restructuring of patterns. Creativity also involves
restructuring but with more emphasis on the escape
from restricting patterns. Lateral thinking involves
restructuring, escape and the provocation of new
patterns.

Lateral thinking is closely related to creativity. But
whereas creativity is too often only the description of a
result lateral thinking is the description of a process.
One can only admire a result but one can learn to use a
process. There is about creativity a mystique of talent
and intangibles. This may be justified in the art world
where creativity involves aesthetic sensibility, emotional
resonance and a gift for expression. But it is not justified
outside that world. More and more creativity is coming
to be valued as the essential ingredient in change and in
progress. It is coming to be valued above knowledge
and above technique since both these are becoming so
accessible. In order to be able to use creativity one must
rid it of this aura of mystique and regard it as a way of
using the mind—a way of handling information. This is
what lateral thinking is about.

Lateral thinking is concerned with the generation of
new ideas. There is a curious notion that new ideas have
to do with technical invention. This is a very minor
aspect of the matter. New ideas are the stuff of change
and progress in every field from science to art, from
politics to personal happiness.

Lateral thinking is also concerned with breaking out of
the concept prisons of old ideas. This leads to changes
in attitude and approach; to looking in a different way at

things which have always been looked at in the same
way. Liberation from old ideas and the stimulation of
new ones are twin aspects of lateral thinking.

Lateral thinking is quite distinct from vertical thinking
which is the traditional type of thinking. In vertical
thinking one moves forward by sequential steps each of
which must be justified. The distinction between the
two sorts of thinking is sharp. For instance in lateral
thinking one uses information not for its own sake but
for its effect. In lateral thinking one may have to be
wrong at some stage in order to achieve a correct
solution; in vertical thinking (logic or mathematics) this
would be impossible. In lateral thinking one may
deliberately seek out irrelevant information; in vertical
thinking one selects out only what is relevant.

Lateral thinking is not a substitute for vertical thinking.
Both are required. They are complementary. Lateral
thinking is generative. Vertical thinking is selective.

With vertical thinking one may reach a conclusion by a
valid series of steps. Because of the soundness of the
steps one is arrogantly certain of the correctness of the
conclusion. But no matter how correct the path may be
the starting point was a matter of perceptual choice
which fashioned the basic concepts used. For instance
perceptual choice tends to create sharp divisions and
use extreme polarization. Vertical thinking would then
work on the concepts produced in this manner. Lateral
thinking is needed to handle the perceptual choice
which is itself beyond the reach of vertical thinking.
Lateral thinking would also temper the arrogance of any
rigid conclusion no matter how soundly it appeared to
have been worked out.

Lateral thinking enhances the effectiveness of vertical
thinking. Vertical thinking develops the ideas generated

by lateral thinking. You cannot dig a hole in a different place by digging the same hole deeper. Vertical thinking is used to dig the same hole deeper. Lateral thinking is used to dig a hole in a different place.

The exclusive emphasis on vertical thinking in the past makes it all the more necessary to teach lateral thinking. It is not just that vertical thinking alone is insufficient for progress but that by itself it can be dangerous.

Like logical thinking lateral thinking is a way of using the mind. It is a habit of mind and an attitude of mind. There are specific techniques that can be used just as there are specific techniques in logical thinking. There is some emphasis on techniques in this book not because they are an important part of lateral thinking but because they are practical. Goodwill and exhortation are not enough to develop skill in lateral thinking. One needs an actual setting in which to practise and some tangible techniques with which to practise. From an understanding of the techniques, and from fluency in their use, lateral thinking develops as an attitude of mind. One can also make practical use of the techniques.

Lateral thinking is not some magic new system. There have always been instances where people have used lateral thinking to produce some result. There have always been people who tended naturally toward lateral thinking. The purpose of this book is to show that lateral thinking is a very basic part of thinking and that one can develop some skill in it. Instead of just hoping for insight and creativity one can use lateral thinking in a deliberate and practical manner.

Summary

The purpose of thinking is to collect information and to make the best possible use of it. Because of the way the

mind works to create fixed concept patterns we cannot
make the best use of new information unless we have
some means for restructuring the old patterns and
bringing them up to date. Our traditional methods of
thinking teach us how to refine such patterns and
establish their validity. But we shall always make less
than the best use of available information unless we
know how to create new patterns and escape from the
dominance of the old ones. Vertical thinking is
concerned with proving or developing concept patterns.
Lateral thinking is concerned with restructuring such
patterns (insight) and provoking new ones (creativity).
Lateral and vertical thinking are complementary. Skill
in both is necessary. Yet the emphasis in education has
always been exclusively on vertical thinking.

The need for lateral thinking arises from the limitations
of the behaviour of mind as a self-maximizing memory
system.

Use of this book

This book is not intended to introduce a new subject nor is it intended to acquaint the reader with what is happening in a certain field. The book is meant to be used. It is meant to be used by the reader for his own sake and through the teacher for the sake of the students.

Age

The processes described in this book are basic ones. They apply to all ages and to all different levels of learning. I have used some of the most elementary demonstrations on the most sophisticated of groups such as advanced computer programmers and they have not felt that they were wasting their time. The more sophisticated the group the better is it able to abstract the process from the particular form in which it is demonstrated. While the lower age groups enjoy the item for its own sake the older age groups look more closely at the point behind it. Although the simpler items are applicable to all age groups the more complicated items may only be of use to more senior groups.

In the younger age groups the visual form is much more effective than the verbal since a child can always attempt to express something visually and, more importantly, to understand something that has been expressed visually.

From the age of seven right up to and through university education the lateral thinking process is relevant. This may seem a wide age group but the process is as basic as logical thinking and clearly the relevance of this is not limited to a particular age group. In a similar manner the relevance of lateral thinking cuts across the distinctions of subject even more than does mathematics. Lateral thinking is relevant whether one is studying science or engineering or history or English. It is because of this general application that the

material used in this book does not require the
background of any particular subject.

An attempt should be made to develop lateral thinking
attitudes as a habit of mind at least from the age of
seven onwards. The actual application of the ideas
expressed in this book to a particular age level must
depend to some extent on the experience of the teacher
in presenting the material in an appropriate form. The
two usual mistakes in this regard are:

● To assume that it is obvious and that everyone
thinks laterally anyway.
● To assume that it is rather a special subject and not
of use or relevance to everyone.

The practical aspect of the book does get more complex
as one proceeds through the book (this is apart from the
background material intended for the teacher). In
general the first part of the practical material is suitable
for seven year olds and the later parts are suitable for
anyone. This is not to imply that the first part is only
suitable for young children or the later parts are only
suitable for adults but that there is a way of putting over
the lateral thinking attitude to any age group.

Format
Like logical thinking, lateral thinking is a general
attitude of mind which may make use of certain
techniques on occasion. Nevertheless this attitude of
mind can best be taught in a formal setting using specific
material and exercises. This is to encourage the
development of the lateral thinking habit. Without a
formal setting one is reduced to mere encouragement
and the appreciation of lateral thinking when it occurs—
neither of which processes do much to develop the
habit.

To set aside a definite period for teaching lateral thinking is much more use than trying to gently introduce its principles in the course of teaching some other subject.

If one has to teach it along with some other subject then one should set aside a short, defined period as part of the general period (even though the subject matter may be the same as for the rest of the period).

A one hour period every week throughout education would be quite sufficient to bring about the lateral thinking attitude – or the creative attitude if you prefer to call it that.

The practical parts of the book are separated into different aspects. It is not suggested that one should work through the book taking a section at each lesson and then passing on to the next section. This would be quite useless. *Instead one uses the basic structure of each section over and again until one is thoroughly familiar with the process. One may spend several sessions on a particular section or even several months.* All the time one is changing the basic material but developing the same lateral thinking process. It is the use of lateral thinking that counts not knowledge of each and every process. One can develop the lateral attitude of mind as easily through thorough practice in one technique as through brief practice in them all.

There is nothing special about the techniques. It is the attitude behind them that counts. But mere exhortation and goodwill are not enough. If one is to develop a skill one must have some formal setting in which to practise it – and some tools to use. The best way to acquire skill in lateral thinking is to acquire skill in the use of a collection of tools which are all used to bring about the same effect.

Materials

Many of the demonstrations used in this book may seem
trivial and artificial. They are. The demonstrations are
used in order to make clear some point about the
thinking process. They are not intended to teach
anything but to encourage the reader to develop some
insight into the natural behaviour of the mind. Just as
the actual content of parables or fables is so much less
important than the point they are intended to convey so
the demonstrations may be trivial in content in order to
make an important point.

There is unfortunately no switch in the mind which can
be flicked one way for dealing with all important matters
and the other way for dealing with minor matters.
Whatever the importance of the matter the system
behaves in the same way, that is according to its nature.
In important matters the working of the system may be
distorted by emotional considerations which do not
interfere with the handling of trivial matters. The only
effect is to make the working worse than it can be.
*Hence the defects of the system in dealing with trivial
matters are at least the same defects which will be
present when dealing with more important matters.*

It is the process not the product that matters. The trivial
and artificial items illustrate the process in a neat and
accessible manner. The process can be extracted just as
the relationships expressed in a formula in algebra can
be separated from what the symbols actually stand for.

Many of the items are visual and even geometric. This is
deliberate because the use of verbal illustrations can be
misleading. Words are already neat and fixed packages
of information and in discussing the thinking process
one really has to go back to the situation itself since the
choice of words in a description is already a choice of
viewpoint, is already quite far along the thinking

process. The nearest one can get to a raw situation, before it has been processed at all by thinking, is a visual situation and geometric ones are preferable since they are more definite and the processing of them is more easily studied. With verbal descriptions quite apart from the choice of viewpoint and the choice of words there are nuances of meaning which can lead to misunderstanding. With a visual situation no meaning is offered. The situation is just there and hence the same for everyone even though they may process it differently.

When the principles indicated by the artificial demonstrations have been understood, when there has been sufficient practice in the processes suggested, then one can move on to more real situations. It is exactly the same as learning mathematics on trivial and artificial problems and then using the processes on important ones.

The amount of material supplied in this book is very limited. What is supplied is supplied more as an example than as anything else. Anyone who is teaching lateral thinking, either to students or to his own children must supplement the material offered here with his own material.

● Visual material
The following material may be collected and used:
1   In the section dealing with the progressive arrangement of cardboard shapes one can make up this sort of shape and also devise new patterns for illustrating the same thing. In addition one can ask the students themselves to devise new shapes.
2   Photographs and pictures can be taken from newspapers and from magazines. These are especially useful in the section on different ways of looking at and interpreting a situation. The captions would naturally

be removed. For convenience the pictures could be mounted on cardboard. If a magazine contained several useful pictures then a number of copies could be bought and used as permanent material.

3   Drawings of scenes or people in action can be provided by the students themselves. A drawing provided by one student is objective material for everyone else. The complexity or accuracy of the drawing is not important since what matters is the way it is looked at by the others.

4   In the sections which call for the execution of designs as drawings these provide abundant material not only for the current set of students but for subsequent ones.

● Verbal material
This can include written, spoken or recorded material.

1   Written material can be obtained from newspapers or magazines.

2   Written material can be supplied by the teacher writing on a particular theme with a definite (even if simulated) point of view.

3   Written material can be supplied by the students who are asked to write a short piece on some particular theme.

4   Spoken material can be derived from radio programmes, from recordings of radio programmes and from deliberate recording of simulated speeches.

5   Spoken material can be obtained from the students themselves one of whom may be asked to talk about a certain subject.

● Problem material
The problem format is a convenient one for encouraging deliberate thinking. It is very difficult to think of a problem just when one is required. There are different sorts of problem.

1   General world problems such as food shortage.

These are obviously open ended problems.

2   More immediate problems such as traffic control in cities. These are problems with which the students may have come into direct contact.

3   Immediate problems. These concern the direct everyday interaction at school. If one does deal with personal problems it is probably best to deal with them in an abstracted way as if talking about third parties.

4   Design and innovation problems. These are requests to bring about a certain effect. They usually apply to concrete objects but they can also apply to organization or ideas (e.g. how would you organize a babysitting service or a supermarket?).

5   Closed problems. These are problems for which there is a definite answer. There is a way of doing something and it is seen to work when it is found. Such problems may be practical ones (for instance how to hang a washing line) or artificial ones (how to make a hole in a postcard big enough to put your head through). Problems can be derived from many different sources:

1   A general glance at a newspaper will generate world or more immediate problems (e.g. strikes).

2   Problems may be suggested by everyday life (e.g. more efficient train services).

3   Problems may be suggested by the students. The teacher asks for problems and then stockpiles the suggestions.

4   Design problems may be generated by taking any item (car, table, desk) and asking how it might be done in a better way. More elaborate design problems can be generated by taking some task which has to be performed by hand and asking for a machine to do the same thing—or a device to make it easier. One could also just ask for a simpler way to do it.

5   Closed problems are rather difficult to find. They must have a definite answer which is difficult enough to make the problem interesting but quite obvious once it has been found. There are some classic problems which

one may know or be told about. It is however a bad idea
to go to a puzzle book since many of the problems
involve quite ordinary mathematical tricks which have
nothing to do with lateral thinking. One simple way of
generating closed problems is to take some ordinary
task and then restrict the starting conditions. For
instance one may want to draw a circle without using a
compass. Once the problem has been set in this way
then one solves it for oneself before offering it to others.

● Themes
There are times when one just wants a subject for
consideration. These are not actual problems nor are
they expressions of a particular point of view. It is a
matter of having a subject area in which to move and
develop ideas (e.g. cups, blackboard, books, acceleration,
freedom, building). These can be obtained in various
ways.
1   Simply by looking around one, taking an object and
elaborating it into a theme.
2   By glancing at a newspaper and deriving a theme for
each headline.
3   By asking the students to generate themes.

● Anecdotes and stories
These are probably the most effective way of putting
across the lateral thinking idea but they are extremely
difficult to generate.
1   From collections of fables or folk stories (e.g.
Aesop's fables, the exploits of the Mulla Nasruddin).
2   By making a note of incidents from one's own
experience or that of others, news items etc.

● Stockpile of material
It always seems much easier to think up material as
required than it really is. It is better to gradually build
up a stockpile of material: newspaper cuttings, photos,

problems, stories, anecdotes, themes and ideas
suggested by the students. One gradually builds up a
file of such things and then can use them as needed. In
addition there is the advantage that with use one can
learn which items are particularly effective. One can
also come to predict the standard responses to the items.
Anecdotes, stories and problems should make a point
about lateral thinking. Themes should be neutral,
specific enough to excite definite ideas but wide enough
for a variety of ideas to be offered. Pictures should be
capable of different interpretations: a man holding a tin
of corned beef is suitable but firemen putting out a fire
is not; a woman looking in a mirror can be ambiguous,
so can policemen arresting a man or soldiers marching
down a street. It is enough if you yourself can think of at
least two different interpretations.

In contrast the verbal material should be as definite as
possible. An article should offer a committed point of
view, even a fanatical point of view. A general
uncommitted appraisal is not so much use unless one is
looking for background information to help
consideration of a theme.

In putting across the idea of lateral thinking as in
teaching any sort of thinking it is possible to talk in
abstract terms but what really makes things clear is
actual involvement. The involvement may start with
abstract geometric shapes and then the process is
transferred bodily to more real situations. It is useful to
keep going back to the simple shapes to emphasize the
process for if one sticks entirely to real situations the
nature of the process may get very blurred. There is also
the real danger that in considering real situations one
comes to think in terms of collecting more information
whereas the whole idea of lateral thinking is concept
restructuring.

Distinctness of lateral thinking
It may seem artificial to separate lateral thinking and
try to teach it on its own when it is so much a part of
thinking. There is a reason for doing this. Many of the
processes of lateral thinking are quite contradictory to
the other processes of thinking (it is their function to be
so). Unless a clear distinction is made there is the danger
of giving the impression that lateral thinking
undermines what is being taught elsewhere by
introducing doubt. It is by keeping lateral thinking
distinct from vertical thinking that one can avoid this
danger and come to appreciate the value of both.
Lateral thinking is not an attack on vertical thinking but
a method of making it more effective by adding
creativity.

The other danger which arises from failure to keep
lateral thinking separate is the vague feeling that one is
teaching it anyway in the course of teaching other things
and therefore there is no need to do anything special
about it. In practice such an attitude is quite wrong.
Everyone naturally feels that they themselves use lateral
thinking and that they always encourage it in their
students. It is very easy to have this feeling but the
fundamental nature of lateral thinking is so different
from that of vertical thinking that it is impossible to
teach both at the same time. It is not enough to
introduce a mild flavour of lateral thinking. One wants
to develop enough skill in it for it to be used effectively
not just acknowledged as a possibility.

Organization of chapters of this book
Each chapter is divided into two parts:
1   Background material, theory and nature of the
process being discussed in that section.
2   Practical format for trying out and using the process
under discussion.

The need for lateral thinking arises from the way the mind works*. Though the information handling system called mind is highly effective it has certain characteristic limitations. These limitations are inseparable from the advantages of the system since both arise directly from the nature of the system. It would be impossible to have the advantages without the disadvantages. Lateral thinking is an attempt to compensate for these disadvantages while one still enjoys the advantages.

Code communication

Communication is the transfer of information. If you want someone to do something you could give him detailed instructions telling him exactly what to do. This would be accurate but it might take rather a long time. It would be much easier if you could simply say to him: 'Go ahead and carry out plan number 4.' This simple sentence might replace pages of instruction. In the military world certain complex patterns of behaviour are coded in this manner so that one only has to specify the code number for the whole pattern of behaviour to be activated. It is the same with computers: much used programmes are stored under a particular heading and one can call them into use by just specifying that heading. When you go into a library to get a book you could describe in detail the book you wanted, giving author, title, subject, general outline etc. Instead of all

---

*A full account of how the mind handles information is given in the book, *The Mechanism of Mind,* published in London by Jonathan Cape (1969) and in New York by Simon and Schuster (1969). It is obviously not possible to cover this matter in detail here for the purpose of this book is different. It is only possible to hint at the type of system involved. Wherever an asterisk occurs in the text (e.g. elsewhere*) those readers who require more detailed information are referred to the other book.

that you could just give the code number from the catalogue.

Communication by code can only work if there are preset patterns. These patterns which may be very complex are worked out beforehand and are available under some code heading. Instead of transferring all the required information you just transfer the code heading. That code heading acts as a trigger word which identifies and calls up the pattern you want. This trigger word can be an actual code heading such as the name of a film or it can be some part of the information which acts to call up the rest. For instance one might not remember a film by its name but if one were to say: 'Do you remember that film with Julie Andrews as a governess looking after some children in Austria?' the rest of the film might be easily brought to mind.

Language itself is the most obvious code system with the words themselves as triggers. There are great advantages in any code system. It is easy to transfer a lot of information very quickly and without much effort. It makes it possible to react appropriately to a situation as soon as the situation is recognized from its code number without having to examine it in detail. It makes it possible to react appropriately to a situation before the situation has even developed fully – by identifying the situation from the initial aspects of it.

It is usual to think of communication as a two way affair: there is someone intending to send a message and someone trying to understand it. An arrangement of flags on a ship's mast is put there intentionally and anyone who understands the code can tell what it means. But a person who knows the code would also be able to pick out a message from a casual arrangement of flags used to decorate a party or a petrol station.

Communication can be a one way business. Dealing with the environment is an example of one way communication. One picks out messages from the environment even though no one has deliberately put them there.

If you offer a random arrangement of lines to a group of people they will soon start to pick out significant patterns. They will be convinced that the patterns have been put there deliberately or that the random arrangements are not random at all but actually constructed out of special patterns. Students who were asked to react in a certain way to a bell which was set off at random intervals soon became convinced that there was a meaningful pattern in the way the bell was sounded.

Communication by code or preset patterns requires the building up of a catalogue of patterns just as you can only use the catalogue number of a book in the library if someone has catalogued the books. As suggested above there does not have to be an actual code number for each pattern. Some part of the pattern itself may come to represent the whole pattern. If you recognized a man by hearing the name 'John Smith' that would be using a code heading, but if you recognized him by the sound of his voice at a party that would be using part of the pattern. Opposite are shown two familiar patterns each of which is partly hidden behind some screen. One would have little difficulty in guessing the patterns from the parts that were accessible.

The mind as a patternmaking system

The mind is a patternmaking system. The information system of the mind acts to create patterns and to recognize them. This behaviour depends on the functional arrangement of the nerve cells of the brain.

The effectiveness of the mind in its one way
communication with the environment arises from this
ability to create patterns, store them and recognize
them. It is possible that a few patterns are built into the
mind and these become manifest as instinctual
behaviour but this seems relatively unimportant in man
as compared to lower animals. The mind can also
accept ready made patterns that are fed to it. But the
most important property of the system is the ability to
create its own patterns. The way the mind actually
creates patterns is described elsewhere*.

A system that can create its own patterns and recognize
them is capable of efficient communication with the
environment. It does not matter whether the patterns
are right or wrong so long as they are definite. Since the
patterns are always artificial ones created by the mind,
it could be said that the function of mind is mistake.
Once the patterns have been formed the selecting
mechanism of usefulness (fear, hunger, thirst, sex, etc)
will sort out the patterns and keep those which are
useful for survival. But first the patterns have to be
formed. The selecting mechanism can only select
patterns; it cannot form them or even alter them.

Self-organizing system
One can think of a secretary actively operating a filing
system, of a librarian actively cataloguing books, of a
computer actively sorting out information. The mind
however does not actively sort out information. The
information sorts itself out and organizes itself into
patterns. The mind is passive. The mind only provides
an opportunity for the information to behave in this
way. The mind provides a special environment in which
information can become self-organizing. This special
environment is a memory surface with special
characteristics.

A memory is anything that happens and does not completely unhappen. The result is some trace which is left. The trace may last for a long time or it may only last for a short time. Information that comes into the
X brain leaves a trace in the altered behaviour of the nerve cells that form the memory surface.

A landscape is a memory surface. The contours of the surface offer an accumulated memory trace of the water that has fallen upon it. The rainfall forms little rivulets which combine into streams and then into rivers. Once the pattern of drainage has been formed then it tends to become ever more permanent since the rain is collected into the drainage channels and tends to make them
X deeper. It is the rainfall that is doing the sculpting and yet it is the response of the surface to the rainfall that is organizing how the rainfall will do its sculpting.

With a landscape the physical properties of the surface will have a strong effect on the way the rainfall affects the surface. The nature of the surface will determine what sort of river is formed. Outcrops of rock will determine which way the river goes.

Instead of a landscape consider a homogeneous surface onto which the rain falls. A shallow dish of table jelly would provide such a surface. If hot water falls on this jelly surface it dissolves a little bit of the jelly and when the water is poured off a shallow depression is left in the surface. If another spoonful of water is poured onto the surface near the first spoonful it will run into the first depression tending to make this deeper but also leaving some impression of its own. If successive spoonfuls of hot water are poured onto the surface (pouring each one off again as soon as it has cooled) the surface will become sculpted into a jelly landscape of hollows and ridges. The homogeneous jelly has simply provided a memory surface for the spoonfuls of hot water to organize themselves into a pattern. The contours of the surface are formed by the water but once formed the contours direct where the water will flow. The eventual pattern depends on *where the spoonfuls of water were placed and in what sequence they were placed*. This is equivalent to the nature of the incoming information and the sequence of arrival. The jelly provides an environment for the self-organization of information into patterns.

Limited attention span

A fundamental feature of a passive self-organizing memory system is the limited attention span. This is why only one spoonful of water at a time was poured onto the jelly surface. The mechanics of how a passive memory surface can come to have a limited attention span are explained elsewhere*. The limited attention span means that only part of the memory surface can be activated at any one time. Which part of the surface comes to be activated depends on what is being presented to the surface at the moment, what has been presented to the surface just before, and the state of the surface (i.e. what has happened to the surface in the past).

This limited attention span is extremely important for it means that the activated area will be a single coherent area and this single coherent area will be found in the most easily activated part of the memory surface. (In the jelly model this would mean the deepest hollow.) The most easily activated area or pattern is the most familiar one, the one which has been encountered most often, the one which has left most trace on the memory surface. And because a familiar pattern tends to be used it becomes ever more familiar. In this way the mind builds up that stock of preset patterns which are the basis of code communication.

With the limited attention span the passive

self-organizing memory surface also becomes a
self-maximizing one. This means that the processes of
selection, rejection, combination and separation all
become possible. Together these processes give the
mind a very powerful computing function*.

Sequence of arrival of information
Overleaf are shown the outlines of two pieces of thin
plastic which are given to someone who is then
instructed to arrange them together to give a shape that
would be easy to describe. The two pieces are usually
arranged to give a square as shown. Then another piece
of plastic is added with the same instructions as before.
This is simply added to the square to give a rectangle.
Two more pieces are now added together. They are put
together to give a slab which is added to the rectangle to
give a square again. Finally another piece is added. But
this new piece will not fit. Although one has been
correct at each stage one is unable to proceed further.
The new piece can not be fitted in to the existing
pattern.

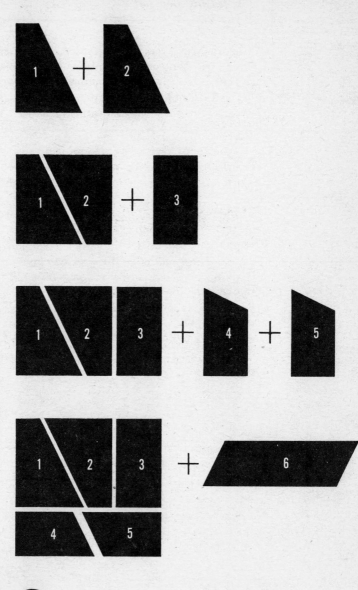

?

A different way of arranging the plastic pieces is shown overleaf. With this new way of arranging them one can fit in all the pieces including the final one. Yet this other method is less likely to be tried than the first method since a square is so much more obvious than a parallelogram.

If one started off with the square then one would have to go back and *rearrange* the pieces at some stage to give a parallelogram before one could proceed. *Thus even though one had been correct at each stage one would still have to restructure the situation before being able to proceed.*

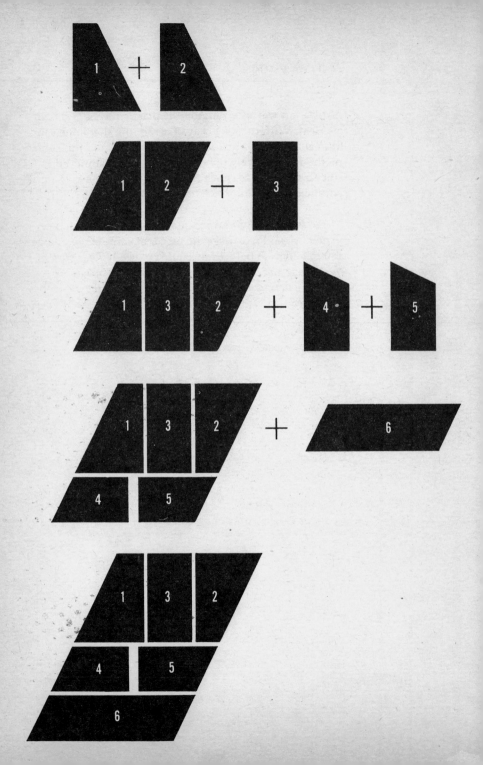

The plastic pieces indicate what happens in a
self-maximizing system. In such a system information
available at any moment is always arranged in the best
way (most stable in physiological terms). As more
information comes in it is added to the existing
arrangement as the plastic pieces were added. But being
able to make sense of the information at several stages
does not mean that one can go on. There comes a time
when one cannot proceed further without restructuring
the pattern – without breaking up the old pattern which
has been so useful and arranging the old information in
a new way.

The trouble with a self-maximizing system that must
make sense at each moment is that the sequence of
arrival of information determines the way it is to be
arranged. For this reason the *arrangement of information
is always less than the best possible arrangement* for the
best possible arrangement would be quite independent
of the sequence of arrival of the pieces of information.

maximum use of
available information

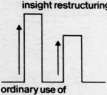

insight restructuring

ordinary use of
available information

In the mind which is a cumulative memory system the
arrangement of information as concepts and ideas tends
to make less than the maximum use of the information
available. This is shown diagrammatically where
the usual level of information use is shown well below
the theoretical maximum level. It is by insight
restructuring that one can move toward the maximal
level.

Humour and insight

As with the plastic pieces there is often an alternative
way of arranging available information. This means that
there can be a switch over to another arrangement.
Usually this switch over is sudden*. If the switch over is
temporary it gives rise to humour. If the switch over is
permanent it gives rise to insight. It is interesting that
the reaction to an insight solution is often laughter even

when there is nothing funny about the solution itself.

A man jumped off the top of a skyscraper. As he passed the third floor window he was heard to mutter: 'So far so good'.
Mr Churchill sat down next to Lady Astor at dinner one day. She turned to him and said, 'Mr Churchill, if I was married to you I should put poison in your coffee.' Mr Churchill turned to her and said, 'Madam, if I was married to you . . . I should drink the coffee.'
A policeman was seen walking along the main street pulling a piece of string. Do you know why he was pulling the piece of string? . . . Have you ever tried *pushing* a piece of string?

In each of these situations an expectation is generated by the way the information is put together. Then suddenly this expectation is thwarted but at once one sees that the unexpected development is another way of putting things together.

Humour and insight are characteristic of this type of information handling system. Both processes are difficult to bring about deliberately.

Disadvantages of the system
The advantages of the preset pattern information system have been mentioned. Basically the advantages are quickness of recognition and hence quickness of reaction. Because one can recognize what one is looking for one can also explore the environment efficiently. The disadvantages are just as definite. Some of the disadvantages of the information handling system of mind are listed here.
1   The patterns tend to become established ever more rigidly since they control attention.
2   It is extremely difficult to change patterns once they have become established.

3   Information that is arranged as part of one pattern cannot easily be used as part of a completely different pattern.

4   There is a tendency towards 'centering' which means that anything which has any resemblance to a standard pattern will be perceived as the standard pattern.

5   Patterns can be created by divisions which are more or less arbitrary. What is continuous may be divided into distinct units which then grow further apart. Once such units are formed they become self-perpetuating. The division may continue long after it has ceased to be useful or the division may intrude into areas where it has no usefulness.

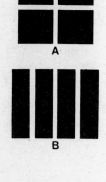

A

B

In the diagram opposite if a square is habitually divided into quarters as shown in A it becomes difficult to use the division shown in B.

6   There is great continuity in the system. A slight divergence at one point can make a huge difference later.

7   The sequence of arrival of information plays too important a part in its arrangement. Any arrangement of information is thus unlikely to be the best possible arrangement of the information that is available.

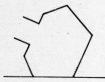

8   There is a tendency to snap from one pattern to another instead of having a smooth change over. This is like those ink bottles which have two stable positions (see opposite). This snapping change occurs as one switches from one stable pattern to another.

9   Even though the choice between two competing patterns may be very fine one of them will be chosen and the other one completely ignored.

10   There is a marked tendency to 'polarize'. This means moving to either extreme instead of maintaining some balanced point between them.

11   Established patterns get larger and larger. That is to say individual patterns are strung together to give a longer and longer sequence which is so dominant that it

constitutes a pattern on its own. There is nothing in the system which tends to break up such long sequences.
12   The mind is a cliché making and cliché using system.

The purpose of lateral thinking is to overcome these limitations by providing a means for restructuring, for escaping from cliché patterns, for putting information together in new ways to give new ideas. In order to do this lateral thinking makes use of the properties of this type of system. For instance the use of random stimulation could only work in a self-maximizing system. Also disruption and provocation are only of use if the information is then snapped together again to give a new pattern.

Summary
The mind handles information in a characteristic way. This way is very effective and it has huge practical advantages. But it also has limitations. In particular the mind is good at establishing concept patterns but not at restructuring them to bring them up to date. It is from these inherent limitations that the need for lateral thinking arises.

Since most people believe that traditional vertical thinking is the only possible form of effective thinking it is useful to indicate the nature of lateral thinking by showing how it differs from vertical thinking. Some of the most outstanding points of difference are indicated below. So used are we to the habits of vertical thinking that some of these points of difference may seem sacrilegious. It may also seem that in some cases there is contradiction for the sake of contradiction. And yet in the context of the behaviour of a self-maximizing memory system lateral thinking not only makes good sense but is also necessary.

Vertical thinking is selective, lateral thinking is generative

Rightness is what matters in vertical thinking. Richness is what matters in lateral thinking. Vertical thinking selects a pathway by excluding other pathways. Lateral thinking does not select but seeks to open up other pathways. With vertical thinking one selects the most promising approach to a problem, the best way of looking at a situation. With lateral thinking one generates as many alternative approaches as one can. With vertical thinking one may look for different approaches until one finds a promising one. With lateral thinking one goes on generating as many approaches as one can even *after* one has found a promising one. With vertical thinking one is trying to select the best approach but with lateral thinking one is generating different approaches for the sake of generating them.

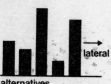

Vertical thinking moves only if there is a direction in which to move, lateral thinking moves in order to generate a direction

With vertical thinking one moves in a clearly defined direction towards the solution of a problem. One uses some definite approach or some definite technique. With lateral thinking one moves for the sake of moving.

One does not have to be moving towards something, one may be moving away from something. It is the movement or change that matters. With lateral thinking one does not move in order to follow a direction but in order to generate one. With vertical thinking one designs an experiment to show some effect. With lateral thinking one designs an experiment in order to provide an opportunity to change one's ideas. With vertical thinking one must always be moving usefully in some direction. With lateral thinking one may play around without any purpose or direction. One may play around with experiments, with models, with notation, with ideas.

The movement and change of lateral thinking is not an end in itself but a way of bringing about repatterning. Once there is movement and change then the maximizing properties of the mind will see to it that something useful happens. The vertical thinker says: 'I know what I am looking for.' The lateral thinker says: 'I am looking but I won't know what I am looking for until I have found it.'

Vertical thinking is analytical, lateral thinking is provocative.

One may consider three different attitudes to the remark of a student who had come to the conclusion: 'Ulysses was a hypocrite.'
1   'You are wrong, Ulysses was not a hypocrite.'
2   'How very interesting, tell me how you reached that conclusion.'
3   'Very well. What happens next. How are you going to go forward from that idea.'

In order to be able to use the provocative qualities of lateral thinking one must also be able to follow up with the selective qualities of vertical thinking.

Vertical thinking is sequential, lateral thinking can make

jumps

With vertical thinking one moves forward one step at a time. Each step arises directly from the preceding step to which it is firmly connected. Once one has reached a conclusion the soundness of that conclusion is proved by the soundness of the steps by which it has been reached.

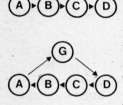

With lateral thinking the steps do not have to be sequential. One may jump ahead to a new point and then fill in the gap afterwards. In the diagram opposite vertical thinking proceeds steadily from A to B to C to D. With lateral thinking one may reach D via G and then having got there may work back to A.

NOTE

When one jumps right to the solution then the soundness of that solution obviously cannot depend on the soundness of the path by which it was reached. Nevertheless the solution may still make sense in its own right without having to depend on the pathway by which it was reached. As with trial-and-error a successful trial is still successful even if there was no good reason for trying it. It may also happen that once one has reached a particular point it becomes possible to construct a sound logical pathway back to the starting point. Once such a pathway has been constructed then it cannot possibly matter from which end it was constructed – and yet it may only have been possible to construct it from the wrong end. It may be necessary to be on the top of a mountain in order to find the best way up.

With vertical thinking one has to be correct at every step, with lateral thinking one does not have to be

NOTE

The very essence of vertical thinking is that one must be right at each step. This is absolutely fundamental to the nature of vertical thinking. Logical thinking and mathematics would not function at all without this

necessity. In lateral thinking however one does not have to be right at each step provided the conclusion is right. It is like building a bridge. The parts do not have to be self-supporting at every stage but when the last part is fitted into place the bridge suddenly becomes self-supporting

With vertical thinking one uses the negative in order to block off certain pathways. With lateral thinking there is no negative

wrong area

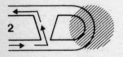

There are times when it may be necessary to be wrong in order to be right at the end. This can happen when one is judged wrong according to the current frame of reference and then is found to be right when the frame of reference itself gets changed. Even if the frame of reference is not changed it may still be useful to go through a wrong area in order to reach a position from which the right pathway can be seen. This is shown diagrammatically opposite. The final pathway cannot of course pass through the wrong area but having gone through this area one may more easily discover the correct pathway.

With vertical thinking one concentrates and excludes what is irrelevant; with lateral thinking one welcomes chance intrusions

Vertical thinking is selection by exclusion. One works within a frame of reference and throws out what is not relevant. With lateral thinking one realizes that a pattern cannot be restructured from within itself but only as the result of some outside influence. So one welcomes outside influences for their provocative action. The more irrelevant such influences are the more chance there is of altering the established pattern. To look only for things that are relevant means perpetuating the current pattern.

With vertical thinking categories, classifications and labels are fixed, with lateral thinking they are not

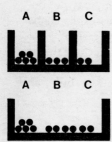

With vertical thinking categories, classifications and labels are useful only if they are consistent for vertical thinking depends on identifying something as a member of some class or excluding it from that class. If something is given a label or put into a class it is supposed to stay there. With lateral thinking labels may change as something is looked at now in one way and now in another. Classifications and categories are not fixed pigeonholes to aid identification but signposts to help movement. With lateral thinking the labels are not permanently attached but are used for temporary convenience.

Vertical thinking depends heavily on the rigidity of definitions just as mathematics does on the unalterable meaning of a symbol once this has been allocated. Just as a sudden change of meaning is the basis of humour so an equal fluidity of meaning is useful for the stimulation of lateral thinking.

Vertical thinking follows the most likely paths, lateral thinking explores the least likely

Lateral thinking can be deliberately perverse. With lateral thinking one tries to look at the least obvious approaches rather than the most likely ones. It is the willingness to explore the least likely pathways that is important for often there can be no other reason for exploring such pathways. At the entrance to an unlikely pathway there is nothing to indicate that it is worth exploring and yet it may lead to something useful. With vertical thinking one moves ahead along the widest pathway which is pointing in the right direction.

Vertical thinking is a finite process, lateral thinking is a probabilistic one

With vertical thinking one expects to come up with an answer. If one uses a mathematical technique an answer is guaranteed. With lateral thinking there may not be any answer at all. Lateral thinking increases the chances for a restructuring of the patterns, for an insight solution. But this may not come about. Vertical thinking promises at least a minimum solution. Lateral thinking increases the chances of a maximum solution but makes no promises.

If there were some black balls in a bag and just one white ball the chances of picking out that white ball would be low. If you went on adding white balls to the bag your chances of picking out a white ball would increase all the time. Yet at no time could you be absolutely certain of picking out a white ball. Lateral thinking increases the chances of bringing about insight restructuring and the better one is at lateral thinking the better are the chances. Lateral thinking is as definite a procedure as putting more white balls into the bag but the outcome is still probabilistic. Yet the pay off from a new idea or an insight restructuring of an old idea can be so huge that it is worth trying lateral thinking for there is nothing to be lost. Where vertical thinking has come up against a blank wall one would have to use lateral thinking even if the chances of success were very low.

Summary
The differences between lateral and vertical thinking are very fundamental. The processes are quite distinct. It is not a matter of one process being more effective than the other for both are necessary. It is a matter of realizing the differences in order to be able to use both effectively.

With vertical thinking one uses information for its own sake in order to move forward to a solution.

With lateral thinking one uses information not for its own sake but provocatively in order to bring about repatterning.

Attitudes towards
lateral thinking 3

Because it is so very different from vertical thinking
many people feel uncomfortable about lateral thinking.
They would rather feel that it is just part of vertical
thinking or that it does not exist. Some of the more
standard attitudes are shown below.

*Although one appreciates the effectiveness of insight
solutions and the value of new ideas there is no practical
way these can be brought about. One can only wait for them
and recognize them after they have happened*

This is a negative attitude which neither takes account
of the insight mechanism nor of the information
imprisoned in cliché patterns. Insight is brought about
by alterations in pattern sequence brought about by
provocative stimulation* and lateral thinking provides
such stimulation. Information imprisoned in old cliché
patterns can often come together in a new way of its own
accord once the pattern is disrupted. It is a function of
lateral thinking to free information by challenging
cliché patterns. To regard insight and innovation as a
matter of chance does not explain why some people are
consistently able to generate more ideas than others. In
any case one can take steps to encourage a chance
process. The effectiveness of lateral thinking for
generating new ideas can be shown experimentally.

*Whenever a solution is said to have been reached by lateral
thinking there is always a logical pathway by which the
solution could have been reached. Hence what is supposed
to be lateral thinking is no more than a plea for better
logical thinking*

It is quite impossible to tell whether a particular
solution was reached by a lateral or vertical process.
Lateral thinking is a description of a process not of a
result. Because a solution could have been reached by
vertical thinking does not mean that it was not reached
by lateral thinking.

If a solution is acceptable at all then by definition there must be a logical reason for accepting it. It is always possible to describe a logical pathway in hindsight *once a solution is spelled out*. But being able to reach that solution by means of this hindsight pathway is another matter. One can demonstrate this quite simply by offering certain problems which are difficult to solve and yet when solved the solution is obvious. In such cases it is impossible to suppose that what made the problem difficult was lack of the elementary logic required.

It is characteristic of insight solutions and new ideas that they should be obvious after they have been found. In itself this shows how insufficient logic is in practice otherwise such simple solutions must have occurred much earlier. In absolute terms it is impossible to prove that a logical pathway could not have been taken if one can be shown in hindsight (except by reference to the mechanics of information handling in the mind). In practical terms however it is quite obvious that the hindsight demonstration of a logical pathway does not indicate that the solution would have been reached in this way.

Since all effective thinking is really logical thinking then lateral thinking is just a part of logical thinking

This objection may seem to be just a semantic quibble. Obviously it does not matter at all whether lateral thinking is regarded as distinct from logical thinking or as part of logical thinking so long as one understands its true nature. If by logical thinking one just means effective thinking then lateral thinking must obviously be included. If by logical thinking one means a sequence of steps each of which must be correct then lateral thinking is clearly distinct.

If the objection takes into account the information

handling behaviour of the mind then it becomes more than a semantic quibble. For in terms of this behaviour it is logical to be illogical. It is reasonable to be unreasonable. If this was not so then I would not be writing a book about it. Here again however one is using logical in terms of 'effective' and not as the operational process we know.

In practice the inclusion of lateral thinking under logical thinking only blurs the distinction and tends to make it unusable – but not unnecessary.

Lateral thinking is the same as inductive logic

This argument is based on the distinction between deductive and inductive logic. The assumption is that anything which is different from deductive logic must be the same as anything else which is also different from deductive logic. There is some resemblance between inductive logic and lateral thinking in that both work from outside the framework instead of from within it. Even so lateral thinking can work from within the framework in order to bring about repatterning by such processes as reversal, distortion, query, turning upside down etc. Inductive logic is essentially reasonable: one tries just as hard to be right as in deductive logic. Lateral thinking however can be deliberately and self-consciously unreasonable in order to provoke a new pattern. Both inductive and deductive logic are concerned with concept forming. Lateral thinking is more concerned with concept breaking, with provocation and disruption in order to allow the mind to restructure patterns.

Lateral thinking is not a deliberate way of thinking at all but a creative gift which some people have and others do not

Some people may be better at lateral thinking just as

some people may be better at mathematics but this does not mean that there is not a process which can be learned and used. It can be shown that lateral thinking can make people generate more ideas and by definition gifts cannot be taught. There is nothing mysterious about lateral thinking. It is a way of handling information.

Lateral thinking and vertical thinking are complementary

Some people are unhappy about lateral thinking because they feel that it threatens the validity of vertical thinking. This is not so at all. The two processes are complementary not antagonistic. Lateral thinking is useful for generating ideas and approaches and vertical thinking is useful for developing them. Lateral thinking enhances the effectiveness of vertical thinking by offering it more to select from. Vertical thinking multiplies the effectiveness of lateral thinking by making good use of the ideas generated.

Most of the time one might be using vertical thinking but when one needs to use lateral thinking then no amount of excellence in vertical thinking will do instead. To persist with vertical thinking when one should be using lateral thinking is dangerous. One needs some skill in both types of thinking.

Lateral thinking is like the reverse gear in a car. One would never try to drive along in reverse gear the whole time. On the other hand one needs to have it and to know how to use it for manoeuvrability and to get out of a blind alley.

Basic nature of lateral thinking 4

In Chapter Two the nature of lateral thinking was indicated by contrasting it with vertical thinking. In this chapter the basic nature of lateral thinking is indicated in its own right.

Lateral thinking is concerned with changing patterns

By pattern is meant the arrangement of information on the memory surface that is mind. A pattern is a repeatable sequence of neural activity. There is no need to define it any more rigidly. In practice a pattern is any repeatable concept, idea, thought, image. A pattern may also refer to a repeatable sequence in time of such concepts or ideas. A pattern may also refer to an arrangement of other patterns which together make up an approach to a problem, a point of view, a way of looking at things. There is no limit to the size of a pattern. The only requirements are that a pattern should be repeatable, recognizable, usable.

Lateral thinking is concerned with changing patterns. Instead of taking a pattern and then developing it as is done in vertical thinking, lateral thinking tries to restructure the pattern by putting things together in a different way. Because the sequence of arrival of information in a self-maximizing system has so powerful an influence on the way it is arranged some sort of restructuring of patterns is necessary in order to make the best use of the information imprisoned within them.

In a self-maximizing system with a memory the arrangement of information must always be less than the best possible arrangement.

The rearrangement of information into another pattern is insight restructuring. The purpose of the rearrangement is to find a better and more effective pattern.

A particular way of looking at things may have
developed gradually. An idea that was very useful at one
time may no longer be so useful today and yet the
current idea has developed directly from that old and
outmoded idea. A pattern may develop in a particular
way because it was derived from the combination of two
other patterns but had all the information been
available at one time the pattern would have been quite
different. A pattern may persist because it is useful and
adequate and yet a restructuring of the pattern could
give rise to something very much better.

In the diagram opposite two pieces come together to
give a pattern. This pattern then combines with another
similar pattern in a straightforward manner. Without
the addition of any new pieces the pattern can suddenly
be restructured to give a much better pattern. Had all
four pieces been presented at once this final pattern is
the one that would have resulted but owing to the
sequence of arrival of the pieces it was the other pattern
that developed.

Lateral thinking is both an attitude and a method of using information

The lateral thinking attitude regards any particular way
of looking at things as useful but not unique or absolute.
That is to say one acknowledges the usefulness of a
pattern but instead of regarding it as inevitable one
regards it as only one way of putting things together.
This attitude challenges the assumption that what is a
convenient pattern at the moment is the only possible
pattern. This attitude tempers the arrogance of rigidity
and dogma. The lateral thinking attitude involves firstly
a refusal to accept rigid patterns and secondly an
attempt to put things together in different ways. With
lateral thinking one is always trying to generate
alternatives, to restructure patterns. It is not a matter of
declaring the current pattern wrong or inadequate.

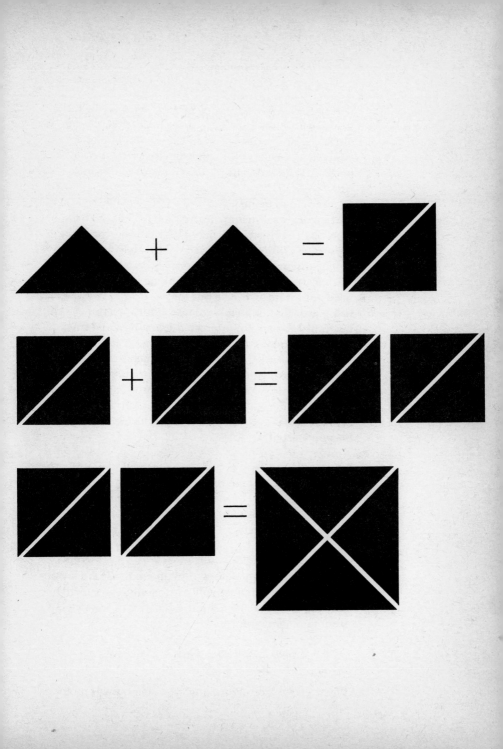

Lateral thinking is never a judgment. One may be quite satisfied with the current pattern and yet try to generate alternative patterns. As far as lateral thinking is concerned the only thing that can be wrong with a pattern is the arrogant rigidity with which it is held.

In addition to being an attitude, lateral thinking is also a particular way of using information in order to bring about pattern restructuring. There are specific techniques which can be used deliberately and these will be discussed later. Underlying them all are certain general principles. In lateral thinking information is used not for its own sake but for its effect. This way of using information involves looking forward not backward: one is not interested in the reasons which lead up to and justify the use of a piece of information but in the effects that might follow such a use. In vertical thinking one assembles information into some structure, bridge or pathway. The information becomes part of the line of development. In lateral thinking information is used to alter the structure but not to become part of it.

One might use a pin to hold two pieces of paper together or one might use a pin to jab into someone and make him jump. Lateral thinking is not stabilizing but provocative. It has to be in order to bring about repatterning. Because it is not possible to restructure a pattern by following the line of development of that pattern lateral thinking may be deliberately perverse. For the same reason lateral thinking may use irrelevant information or it may involve suspending judgment and allowing an idea to develop instead of shutting it off by pronouncing it wrong.

Lateral thinking is directly related to the information handling behaviour of mind

The need for lateral thinking arises from the limitations

NOTE of a self-maximizing memory system. Such a system functions to create patterns and then to perpetuate them. The system contains no adequate mechanism for changing patterns and bringing them up to date. Lateral thinking is an attempt to bring about this restructuring or insight function.

Not only does the need for lateral thinking arise from the information handling of mind but the effectiveness of lateral thinking also depends on this behaviour. Lateral thinking uses information provocatively. Lateral thinking breaks down old patterns in order to liberate information. Lateral thinking stimulates new pattern formation by juxtaposing unlikely information. All these manoeuvres will only produce a useful effect in a self-maximizing memory system which snaps the information together again into a new pattern. Without this behaviour of the system lateral thinking would be purely disruptive and useless.

Once one has acquired the lateral thinking attitude one does not need to be told on what occasions to use lateral thinking.

Throughout this book lateral thinking is kept quite distinct from vertical thinking in order to avoid confusion. This is also done so that one can acquire some skill in lateral thinking without impairing one's skill in vertical thinking. When one is thoroughly familiar with lateral thinking one no longer has to keep it separate. One no longer has to be conscious whether one is using lateral or vertical thinking. The two blend together so that at one moment vertical thinking is being used and the next moment lateral thinking is being used. Nevertheless there are certain occasions which call for the deliberate use of lateral thinking.

New Ideas

Most of the time one is not conscious of the need for new ideas even though one is grateful enough when they turn up. One does not try and generate new ideas because one suspects that new ideas can not be generated by trying. Though new ideas are always useful there are times when one is very much aware of the need for a new idea. There are also jobs which demand a continual flow of new ideas (research, design, architecture, engineering, advertising etc).

The deliberate generation of new ideas is always difficult. Vertical thinking is not much help otherwise new ideas would be far easier to come by, indeed one would be able to programme a computer to churn them out. One can wait for chance or inspiration or one can pray for the gift of creativity. Lateral thinking is a rather more deliberate way of setting about it.

Many people suppose that new ideas mean new inventions in the form of mechanical contrivances. This

is perhaps the most obvious form a new idea can take
but new ideas include new ways of doing things, new
ways of looking at things, new ways of organizing
things, new ways of presenting things, new ideas about
ideas. From advertising to engineering from art to
mathematics, from cooking to sport, new ideas are
always in demand. This demand need not be just a
general inclination but can be as specific as one likes.
One can actually set out to generate new ideas.

Problem solving

Even if one has no incentive to generate new ideas
problems are thrust upon one. There is little choice but
to try and solve them. A problem does not have to be
presented in a formal manner nor is it a matter for
pencil and paper working out. *A problem is simply the
difference between what one has and what one wants.* It
may be a matter of avoiding something, of getting
something, of getting rid of something, of getting to
know what one wants.

There are three types of problem:

- The first type of problem requires for its solution more
  information or better techniques for handling
  information.
- The second type of problem requires no new
  information but a rearrangement of information already
  available: an insight restructuring.
- The third type of problem is the problem of no
  problem. One is blocked by the adequacy of the present
  arrangement from moving to a much better one. There
  is no point at which one can focus one's efforts to reach
  the better arrangement because one is not even aware
  that there is a better arrangement. The problem is to
  realize that there is a problem – to realize that things can
  be improved and to define this realization as a problem.

The first type of problem can be solved by vertical
thinking. The second and third type of problem require

lateral thinking for their solution.

Processing perceptual choice

Logical thinking and mathematics are both second stage
information processing techniques. They can only be
used at the end of the first stage. In this first stage
information is parcelled up by perceptual choice into
the packages that are so efficiently handled by the
second stage techniques. It is perceptual choice which
determines what goes into each package. *Perceptual
choice is the natural patterning behaviour of mind.*
Instead of accepting the packages provided by
perceptual choice and going ahead with logical or
mathematical processing one might want to process the
packages themselves. To do this one would have to use
lateral thinking.

Periodic reassessment

Periodic reassessment means looking again at things
which are taken for granted, things which seem beyond
doubt. Periodic reassessment means challenging all
assumptions. It is not a matter of reassessing something
because there is a need to reassess it; there may be no
need at all. It is a matter of reassessing something
simply because it is there and has not been assessed for a
long time. It is a deliberate *and quite unjustified* attempt
to look at things in a new way.

Prevention of sharp divisions and polarizations

Perhaps the most necessary use of lateral thinking is
when it is not used deliberately at all but acts as an
attitude. As an attitude lateral thinking should prevent
the emergence of those problems which are only created
by those sharp divisions and polarizations which the
mind imposes on what it studies. While acknowledging
the usefulness of the patterns created by mind one uses
lateral thinking to counter arrogance and rigidity.

Techniques

The preceding chapters have dealt with the nature and use of lateral thinking. In reading through them one may have developed a clear idea of what lateral thinking is about. The more usual reaction is to understand and accept what has been written as one reads it and then to forget about it so quickly that one only retains a vague impression of what lateral thinking is about. Nor is this surprising because ideas are insubstantial things. Even if one did obtain a clear idea of the nature of lateral thinking it would be very difficult to pass on this idea without incorporating it in something more substantial.

A nodding acknowledgement of the purpose of lateral thinking is not much good. One has to develop some skill in the actual use of this type of thinking. Such skill can only develop if one has enough practice. Such practice ought not to await formal organization but it very often does. The techniques that are outlined in the following pages are meant to provide formal opportunities for practising lateral thinking. Some of the techniques may seem more lateral than others. Some of them may even seem to be things one always does anyway – or at least always imagines that one does.

Underlying each of these techniques are the basic principles of the lateral use of information. One does not have to stress these or lay them bare.

The purpose of the formal techniques is to provide an opportunity for the practical use of lateral thinking so that one may gradually acquire the lateral thinking habit. The techniques are not suggested as formal routines which must be exactly learned so that they can be deliberately applied thereafter. Nevertheless the techniques can be used in this manner and until one acquires sufficient fluency in lateral thinking to do without formal techniques one can use them as such.

Each section is divided into two parts. The first part is
concerned with the nature and purpose of the technique.
The second part consists of suggestions for the actual
practice of the technique in a classroom or other setting.
The material offered is only meant to suggest the sort of
material that a teacher might assemble. The collection
of further material and the handling of the practice
sessions was discussed in the special section at the
beginning of this book.

vertical

lateral

The most basic principle of lateral thinking is that any particular way of looking at things is only one from among many other possible ways. Lateral thinking is concerned with exploring these other ways by restructuring and rearranging the information that is available. The very word 'lateral' suggests the movement sideways to generate alternative patterns instead of moving straight ahead with the development of one particular pattern. This is indicated in the diagrams opposite.

It may seem that the search for alternative ways of looking at something is a natural search. Many people feel that this is something that they always do. To some extent it is but the lateral search for alternatives goes far beyond the natural search.

In the natural search for alternatives one is looking for the best possible approach, in the lateral search for alternatives one is trying to produce as many alternatives as possible. One is not looking for the *best* approach but for as many *different* approaches as possible.

In the natural search for alternatives one stops when one comes to a promising approach. In the lateral search for alternatives one acknowledges the promising approach and may return to it later but one goes on generating other alternatives.

In the natural search for alternatives one considers only reasonable alternatives. In the lateral search for alternatives these do not have to be reasonable.

The natural search for alternatives is more often an intention than a fact. The lateral search for alternatives is deliberate.

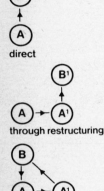

direct

through restructuring

indirect

The main difference is the purpose behind the search for alternatives. The natural inclination is to search for alternatives in order to find the best one. In lateral thinking however the purpose of the search is to loosen up rigid patterns and to provoke new patterns. Several things may happen with this search for alternatives. One may generate a number of alternatives and then return to the original most obvious one.
A generated alternative might prove a useful starting point.
A generated alternative might actually solve the problems without further effort.
A generated alternative might serve to rearrange things so that the problem is solved indirectly.

Even if the search for alternatives proves to be a waste of time in a particular case it helps develop the habit of looking for alternatives instead of blindly accepting the most obvious approach.

The search for alternatives in no way prevents one from using the most obvious approach. The search merely delays the use of the most probable approach. The search merely adds a list of alternatives to the most probable approach but detracts nothing from it. In fact the search adds to the value of the most probable approach. Instead of this approach being chosen because it seems the only one, it is chosen because it is obviously the best from among many other possibilities.

Quota
In order to change the search for alternatives from being a good intention to a practical routine one can set a quota. A quota is a fixed number of alternative ways of looking at a situation. The advantage of having a predetermined quota is that one goes on generating alternatives until one has filled the quota and this means that if a particularly promising alternative occurs early

in the search one acknowledges it and moves on instead
of being captured by it. A further advantage of the
quota is that one has to make an effort to find or
generate alternatives instead of simply awaiting the
natural alternatives. One makes an effort to fill the quota
even if the alternatives generated seem artificial or even
ridiculous. Suitable quotas might be three, four or five
alternatives.

Having a quota does not of course stop one generating
even more alternatives but it does ensure that one
generates at least the minimum.

Practice
Geometric figures
The advantage of visual figures is that the material is
presented in an unequivocal form. A student may look
at the material and make of it what he will but the
material remains the same. This is in contrast to verbal
material where tone, emphasis, individual shades of
meaning all give the material an individual flavour
which is not available to everyone.

The advantage of geometrical figures is that they are
standard patterns described by simple words. This
means that one can snap from one description to
another without any difficulty in describing how one is
looking at the figure.

The teacher starts off with the geometric figures in order
to indicate what the generation of alternatives is all
about. When the idea is clear he can move on to less
artificial situations.

In practice the teacher handles the situation as follows:
1   The figure is shown on the board to the whole class
or else given out to each student on a separate piece of
paper.

figure

A
a triangle sitting
on a rectangle

B
a square with
two upper
corners missing

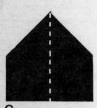

C
two halves of
a rectangle
put side by side

D
end view of
a house

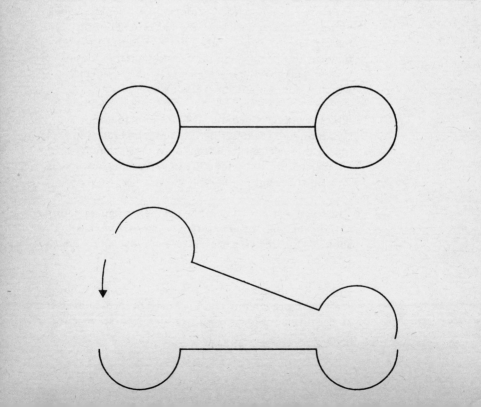

2   The students are asked to generate different ways of describing the figure.

3   The teacher can then collect the written alternatives or not, depending on the size of the classroom and the available time.

4a   (papers not collected)
The teacher asks for a volunteer description of the figure. If one is not forthcoming he points at someone and asks that person to describe the figure. Having got the first description the teacher asks for other variations. The other possible variations are listed.

4b   (papers collected)
The teacher may pick out one or two papers without needing to go through the lot. He reads out the description. He then asks for other variations or goes through the accumulated papers and picks out any variations.

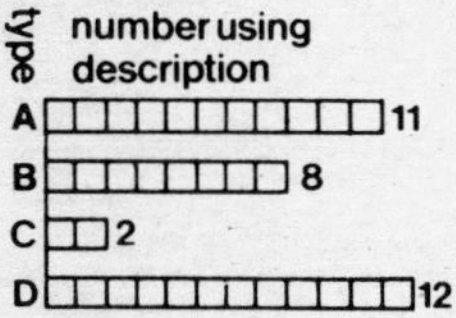

If there is sufficient time between sessions the teacher could go through the papers and draw up a histogram list of the variations offered (as shown opposite). This is then shown at a subsequent session.

type

number using
description

A [ ] 11
B [ ] 8
C [ ] 2
D [ ] 12

5   The function of the teacher is to encourage and accept variations not to judge them. If a particular variation seems outrageous the teacher does not condemn it but asks the originator to explain it more fully. If it is obvious that the rest of the classroom cannot be persuaded to accept this outrageous variation then it is best to list it at the bottom. But it should not be rejected.

6   Whenever there is difficulty in generating variations the teacher must insert a few possibilities which he himself has prepared beforehand.

Material
1   How would you describe the figure shown opposite?
Alternatives
Two circles joined by a line.
A line with a circle at either end.

Two circles each with a short tail attached and placed so
that the tails are in line and meet up.
Two pieces of guttering, one placed on top of the other.
Comment
It may be protested that 'two circles joined by a line' is
really the same as a 'line with a circle at either end'. This
is not so since in one case attention starts with the circle
and in the other case it starts with the line. From the
point of view of what happens in the mind the sequence
of attention is of the utmost importance hence a
different sequence of attention is a difference.

Some of the descriptions may be static ones that can be
explained in terms of the figure shown. Others may be
dynamic descriptions which are more easily shown by
additional diagrams. This happens when the presented
diagram is taken as the end point of some arrangement
of other figures.

2　　How would you describe the figure shown opposite?
Alternatives
An L shape.
A carpenter's angle.
A gallows upside down.
Half a picture frame.
Two rectangles placed one against the other.
A large rectangle with a smaller rectangle subtracted.

Comment
Some difficulty arises when the presented shape is
compared to an actual object like 'a carpenter's angle'.
The difficulty is that this sort of description opens up an
unlimited range of descriptions, for instance another
description might describe the shape as a building
looked at from the air. The point to keep very clearly in
mind is that *one is asked for an alternative description of
the presented figure, one is not asking what the figure could
be or what it reminds you of.* The description must be

such that someone could actually draw the figure from
the description. Thus the suggestion that the figure
looks like a building seen from the air is useless unless
the building is specified as L shaped in which case the
description is L shaped. One need not insist that the
description be very exact, for instance the 'two
rectangles placed against the other' ought really to
contain an indication of the orientation but one must
not be pedantic because it misplaces the emphasis.

Some of the descriptions may indicate a particular
process. Such descriptions as 'two rectangles placed one
against the other' or 'a larger rectangle with a smaller
rectangle missing' actually require that one consider

some other figure and then subtract or modify. Clearly
this is a valid method of description. The basic types of
description might be regarded as:
Building up from smaller units.
Comparing to another figure.
Modifying another figure by addition or subtraction.

As before one may have to draw additional diagrams to
show what is meant. If one cannot understand oneself
what the student means then he is asked to explain it
himself.

3   How would you describe the figure shown opposite?
Alternatives
Two overlapping squares.
Three squares.
Two L shapes embracing a square gap.
A rectangle divided into half with the two pieces pushed
out of line.
Comment
The 'two overlapping squares' seems so obvious a
description that any other seems perverse. This
illustrates how strong is the domination by obvious
patterns. Once again it may be felt that 'two squares
overlapping' is the same as 'three squares' since the
latter is implied by the former. This is a tendency that
must be resisted because often even a minor change in
the way a thing is looked at can make a huge difference.
One must resist the temptation to say that one
description means the same thing as another and hence
that it is just quibbling.

There may be elaborate descriptions which seek to be so
comprehensive that they cover all possibilities:
'Two squares that overlap at one corner so that the area
of overlap is a square of side about half that of the
original squares'. Such comprehensive descriptions
almost reproduce the diagram and hence must include

all sorts of other descriptions. Nevertheless these other
descriptions must be accepted in their own right.
Logically a description may be redundant in that it is
implied by another but perceptually the same
description may make use of new patterns. For instance
the idea of *three* squares is useful even though it is
implicit in the overlap description.

4   How is the pattern opposite made up?
Alternatives
A small square surrounded by big squares.
A big square with small squares at the corners.
A column of large squares pushed sideways to give a
staircase pattern.
Basic unit made out of one large and one small square.
Extend the edges of a small square and draw other small
squares on these extended edges.
A line is divided into thirds and perpendiculars are
drawn at each third.
In a grid pattern some of the small squares are
designated in a certain way and outlined and then the
lines are removed and the spaces filled with big squares.
Big squares are placed against each other so that the side
of each one half overlaps the side of every adjacent
square.
Two overlapped patterns of lines, one at right angles to
the other.
Comment
There are very many possible variations other than
those listed above. The descriptions offered must be
workable. The description should clearly indicate how
the pattern is being looked at. What is of importance is
the variety of ways the pattern can be treated: in terms
of large squares only, in terms of small squares only,
in terms of both large and small squares, in terms of
lines, in terms of spaces, in terms of a grid pattern.

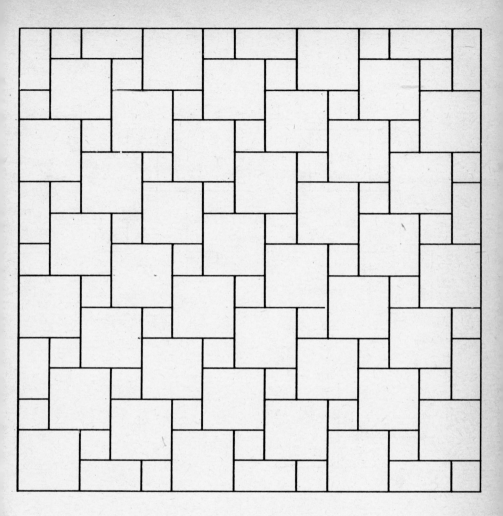

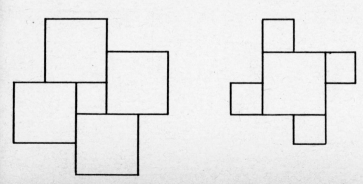

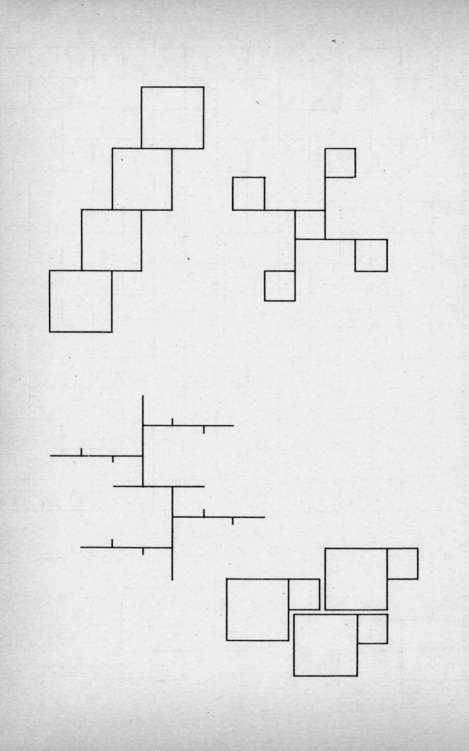

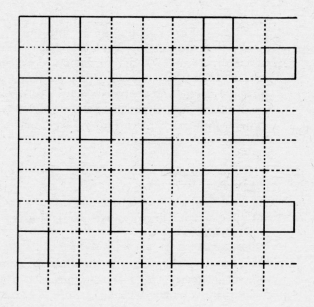

● Activity
The examples used so far call for different descriptions
of a presented pattern. One can move on from different
ways of looking at things to different ways of doing
things. This is rather more difficult since with
description it is only a matter of selecting what is
already there but to do something one has to put in
what is not there.

5   How would you divide a square into four equal
pieces? (For this example it is better that each student
tries to draw as many different versions as he can
instead of just watching the board and offering a new
approach. At the end the papers may be collected if the
teacher wants to analyse the results or else left with the
students for them to tick off the various versions.)
Alternatives
Slices.
Four smaller squares.
Diagonals.
Divide the square into sixteen small squares and then
put these together to give swastika or L shapes as shown.
Other shapes as shown.
Comment
Many students at first stick to the slices, diagonals and
four small squares. One then introduces the idea of
dividing the square into sixteen small squares and
putting these together in different ways. The next
principle is that any line which passes from a point on
the edge of the square to an equivalent point on the
opposite edge and has the same shape above the centre
point as below it divides the square into half. By
repeating the line at right angles one can divide the
square into quarters. Obviously there is an infinite
number of shapes which this line can have. It may be
that some students will offer variations on this principle
without realizing the principle. Rather than listing each

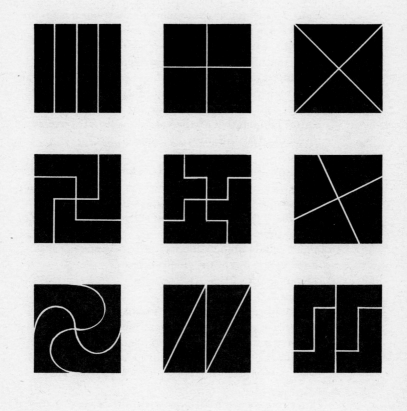

variation one puts them together under the one
principle. A variation on this principle involves dividing
the square into half and then dividing each half into half
again. For each half any division which passes through
the centre of that half and is of equivalent shape on each
side of the centre point will do. This introduces a whole
new range of shapes.
Since this is not an exercise in geometry or design the
intention is not to explore the total possible ways of
carrying out the division. What one tries to do is to
show that there are other ways even when one is
convinced that there cannot be. Thus the teacher waits
until no further ways are offered and then introduces
the variations suggested above one at a time. (It may of
course happen that all the variations listed above are
introduced by the students themselves.)

6   How would you divide up a square of cardboard to
give an L shape with the same area as the square. You
can use not more than two cuts. (Actual squares of
cardboard can be used or drawings should suffice.)
Alternatives
The two rectangular slices (see figure opposite.)
The cutting out of the small square.
The diagonal cut.
Comment
The requirement 'use not more than two cuts'
introduces the element of constraint. The constraint is
not meant to be restrictive, on the contrary it
encourages the effort to find difficult alternatives instead
of being easily satisfied.

Since one is used to dealing with vertical and horizontal
lines and with right angles the diagonal method is not
easy to find. Perhaps the best way to find it is to 'cut
across the square diagonally and then see where that
gets one'. In effect one is beginning to use provocative
manoeuvres rather than simple analytical ones.

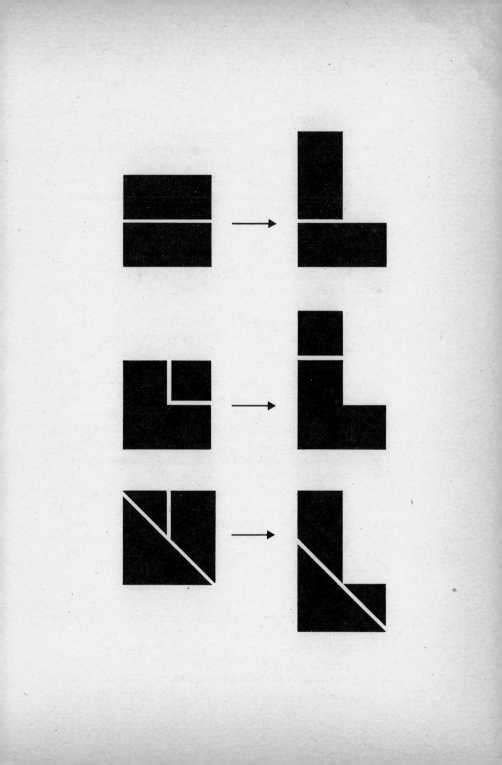

Nongeometrical shapes
Having used the geometrical shapes to illustrate the
deliberate search for alternatives (and also the possibility
of such alternatives) one can move on to more complex
situations. In these more complex situations it is not so
much a matter of picking out standard patterns as
alternatives but of putting things together to give a
pattern.

7   A one pint milk bottle with half a pint of water in it.
How would you describe that bottle?
Alternatives
A half empty bottle of water.
A milk bottle half filled with water.
Half a pint of water in an empty one pint milk bottle.
Comment
In itself the milk bottle example is trivial. But it does
serve to illustrate how there can be two completely
different ways of looking at something. It also shows
that when one way has been chosen the alternative way
is usually ignored. It is of interest that when the bottle is
half filled with milk it is more often described as half
empty, but when it is half filled with water it tends to be
described as half full. This probably happens because
in the case of the milk one is working downward from a
full bottle but in the case of the water one is working
upward from an empty milk bottle. The history of a
situation has much effect on the way it is looked at.

● Pictures
Photographs from newspapers or magazines are the
most easily available source of pictures. The difficulty is
to make them available to a large group. This could be
done by getting individual copies of a newspaper and
keeping them until the material is out of date. If
sufficiently skilled the teacher could actually draw
pictures on the board but this is much less satisfactory.
The type of material needed has been discussed in the

section 'The Use of this book'.

Pictures can be used in two ways
- Describe what you think is happening in that picture.
- Describe three different things that could be happening in that picture.

In method 1 the teacher uses an ambiguous picture and asks each person to make his own interpretation. At the end he collects the interpretations. Variability between individual interpretations shows the alternative ways of looking at the picture. The teacher is careful not to judge which way is best or why one way is unreasonable. Nor does he reveal what the picture was actually about (he can conveniently have forgotten this).

In method 2 the students are asked to generate a quota of different interpretations. If the students tend to be blocked by the most obvious interpretation and are unwilling to guess at any others then they may be allowed to list the interpretations in order of likelihood. In addition the teacher throws in some outlandish suggestions about the particular picture being used in order to suggest what is required.

Examples
A photograph showing a group of people wading through shallow water. They are not dressed for paddling. In the background appears to be a beach. The following interpretations were received:
A group of people caught by the tide.
People crossing a flooded river.
People wading out to an island or sand spit.
Wading through flood water.
People wading out to a ferry boat which cannot come inshore.
People coming ashore from a wrecked boat.

Comment
In fact the photograph showed a group of people
protesting at the poor state of the beach. It was not
important that anyone should have guessed this since it
was not an exercise in logical deduction. What was
important was that there were several different
interpretations of what was going on. Apart from noting
these variations one should have been able to generate
them (even if only to reject them).

Example
Photograph of a boy sitting on a park bench.
Alternatives
Picture of an inactive or lazy boy.
An empty space on a park bench.
Part of the bench is being kept dry by the boy.
Comment
The description of this picture is quite different from
the other example. There is less attempt to say what is
happening (e.g. a boy waiting for his pals, a tired boy
resting, a boy playing truant from school, a boy
enjoying the sun). Instead the description is directed at
the scene itself rather than the meaning (e.g. a boy on a
park bench, an empty space on the bench). There is also
an attempt to look at the picture in an unusual way.
This might have gone too far with 'part of the bench
being kept dry by the boy' but there really are no limits.
In any picture there are several different levels of
description: what is shown, what is going on, what
has happened, what is about to happen. In asking for
alternatives the teacher may leave it quite open at first
but later on he specifies the level of description within
which the alternatives have to be generated.

● Altered pictures
The trouble with pictures is that too often the obvious
interpretation is completely dominant. Not only is it
difficult to find other ways of looking at it but these

other ways seem silly and artificial. To avoid this difficulty and to make things more interesting the teacher can alter pictures by covering up parts of them. It immediately becomes far more difficult to tell what the picture is about from the exposed part and thus one is able to generate alternative possibilities without being dominated by an obvious interpretation. There is also the added incentive of trying to guess the right answer which will be obvious when the full picture is revealed.

Example
Half of a picture is obscured. What is revealed is a man balancing on the edge of a ledge running along the side of some building.
Alternatives
A man threatening to commit suicide.
Rescuing a cat that has got stuck on a ledge.
Escaping from a burning building.
Film stunt man.
A man trying to get into his room having locked himself out.
Comment
The rest of the picture would have shown some student posters which the man was sticking up. The use of partial pictures makes it easier to generate alternatives but ultimately one wants to be able to restructure pictures in which an obvious interpretation makes it difficult to find alternative structurings. It is especially those situations which are dominated by an obvious interpretation that one wants to practice restructuring. One can use the easier partial pictures however to acquire experience. Another advantage of the partial picture is that it indicates that the interpretation may lie outside what is visible. This makes one inclined to look about not only at what is in the actual situation being examined but at things outside it.

● Written material—Stories
Stories may be obtained from newspapers or magazines
or even from books that are being used elsewhere in the
curriculum. By story is not meant a tale but any written
account.

Stories may be treated in the following ways:
● Generate the different points of view of the people
involved.
● Change what is a favourable description to an
unfavourable one not by changing the material but by
changing the emphasis and looking at it in a different
way.
● Extract a different significance from the information
given than that extracted by the writer.

Example
Newspaper story of an eagle that has escaped from the
zoo and is proving difficult to capture. It is perched on a
high branch and is resisting the efforts of the keepers to
lure it back to its cage.
Alternatives
The keeper's point of view: the bird may fly away and
get lost or shot unless it is coaxed back soon. It is
uncomfortable having to climb up trees after the bird
and one feels a bit of a fool. Someone is to blame for
having let it escape.
The newspaperman's point of view: the longer the bird
stays out the better the story. Can one get close enough
to get a good picture. One ought to find some other
interest such as different people's ideas on how to catch
the bird.
The eagle's point of view: wondering what all the fuss is
about. Strange feeling not to be in a cage. Getting
rather hungry. Not sure in which direction to fly.
The onlooker's point of view: hoping the eagle will fly
away and be free for evermore. Amused to see the
strenuous efforts being made to catch the bird. The

eagle looks so much better out on its own than inside a cage. Perhaps one could show how clever one was by catching the bird when no one else could.

Comment
Whenever there is a story with different people involved then it is a simple matter to try to generate the point of view of everyone concerned. Every student could try to generate the different points of view or else different students could be assigned to generate the different points of view. The exercise is not so much to try to guess what other people are thinking but to show how the same situation can be structured in different ways.

Example
A story describing the uncomfortable life in a primitive community where the people cannot read or write and where only a bare subsistence can be obtained by hard work in the fields.

Alternatives
Comfort as a matter of what one was used to.
If one was used to simple things and could obtain simple things perhaps this was better than expecting complex things and being dissatisfied when one could not obtain them.
Perhaps reading and writing only upset people by making them aware of the awful things that are happening in the rest of the world.
Perhaps reading and writing make people more dissatisfied.
Most people are usually working hard at something or other, perhaps hard work in the field is more rewarding since one can actually see something growing and one is actually going to eat what one grows.

Comment
The alternative point of view does not necessarily have to be the point of view held by the person generating it. The person may actually hold exactly the same point of view as the writer. The purpose is to show that one can

look at things in a different way. Nor is it a matter of
trying to prove one point of view to be better than the
other. There is no question of arguing for instance 'that
the simple community may seem pleasant but if one is
ill one must just die etc.' In practice it is difficult to
avoid arguing. It is also difficult to put forward a point
of view with which one does not agree. The advantage
of being able to put forward an opposing point of view is
that one then has much more chance of restructuring it.

Example
A story may cite the long hair and colourful clothes of
young men as an example that they were being
demasculinized and becoming effeminate; that one
could no longer distinguish between boys and girls.
Alternative
Wearing long hair shows courage, it shows the courage
to defy conventions.
Until quite recently men always wore long hair as in the
Elizabethan era and far from being less masculine they
were more masculine. As for the colourful clothes these
were flamboyant not feminine. They indicated a
masculine search for individuality.
In any case why shouldn't boys and girls look alike.
In that way at least girls would get equal rights.
Comment
In this type of restructuring no extra information may
be introduced. It is definitely not meant to be a
presentation of the other side of the case. The purpose
is to show that the material put together to give one
point of view can also be put together in a completely
different way.

Problems
Problems can be generated from the inconveniences of
everyday living or by looking through a newspaper.
Newspaper columns are full of difficulties, disturbances,
things that have gone wrong and complaints. Though

these may not actually be stated as problems they can easily be rephrased as such. It is enough that a general problem theme be stated; there is no need to set up a formal problem. Any situation where there is room for improvement can be used as a problem and also any difficulty that can be imagined.

In using problem material to exercise the generation of alternatives one may proceed in two ways:
1   Generate alternative ways of stating the problem.
2   Generate alternative approaches to the problem.
The emphasis is not on actually trying to solve the problem but on finding different ways of looking at the problem situation. One may go on towards a solution but this is not essential.

Example
The problem of children getting separated from their parents in large crowds.
Alternatives
1   Restatements
Preventing separation of children from parents.
Preventing children being lost.
Finding or returning lost children.
Making it unnecessary for parents to have to take children into large crowds, (crèches at exhibitions etc).
Comment
Some of the alternative statements of the problem do suggest answers. The more general the statement of a problem the less likely is it to suggest answers. If a problem is stated in very general terms then it is not easy to restate the problem in another way on the same level of generality. If this is the case one can always descend to a more specific level in order to generate alternatives. For instance 'the problem of lost children in crowds' could be restated as the 'problem of careless parents in crowds' or 'the problem of children in crowds' but one could also use a more specific level such as 'the problem of returning lost children to their parents'.

2   Different approaches
Alternatives
Attach children more firmly to their parents (by a dog's lead?).
Better identification of children (disc with address).
Make it unnecessary for children to be taken into the crowd (crèches etc).
Central points for children and parents to get to if losing sight of one another.
Display list of lost children.
Comment
In this case many of the approaches seem like actual solutions. In other situations however approaches may just indicate a way of tackling the problem. For instance with this lost children problem one approach might be 'collect statistical data on how many people take their children into crowds because they want the children to be there or because there is no one the children can be left with.'

● Type of problem
The type of problem used depends very much on the age of the students involved. The problem suggestions listed below are divided into a young age group and an older one.
Young age group
Making washing up easier or quicker.
Getting to school on time.
Making bigger icecreams.
Getting a ball that is stuck in a tree.
How to manage change on buses.
Better umbrellas.
Older age group
Traffic jams.
Room for airports.
Making railways pay.
Enough low cost housing.
World food problem.

What should cricketers do in winter.
Better design for a tent.

Summary
This chapter has been concerned with the deliberate
generation of alternatives. This generation of
alternatives is for its own sake and not as a search for the
best way of looking at things. The best way may
become obvious in the course of the procedure but one
is not actually trying to find it. If one was just looking
for the best approach then one would stop as soon as
one found what appeared to be the best approach.
Instead of stopping however one goes on with the
generation of alternatives for its own sake. The purpose
of the procedure is to loosen up rigid ways of looking at
things, to show that alternative ways are always present
if one bothers to look for them, and to acquire the habit
of restructuring patterns.

It is probably better to use the artificial quota method
rather than just rely on the general intention for trying
to find other ways of looking at things. General
intentions work well when things are easy but not when
they are difficult. The quota sets a limit which *must* be
met.

The previous chapter was concerned with alternative ways of putting things together. It was a matter of finding out alternative ways of putting A, B, C and D together to give different patterns. This section is concerned with A, B, C and D for their own sakes. Each of them is itself an accepted, standard pattern.

A cliché is a stereotyped phrase, a stereotyped way of looking at something or describing something. But clichés refer not only to arrangements of ideas but to ideas themselves. It is usually assumed that the basic ideas are sound and then one starts fitting them together to give different patterns. But the basic ideas are themselves patterns that can be restructured. It is the purpose of lateral thinking to challenge any assumption for it is the purpose of lateral thinking to try and restructure any pattern. General agreement about an assumption is no guarantee that it is correct. *It is historical continuity that maintains most assumptions – not a repeated assessment of their validity*.

The figure overleaf shows three shapes. Suppose you had to arrange them to give a single shape that would be easy to describe? There is difficulty in finding such an arrangement. But if instead of trying to fit the given shapes together one reexamined each shape then one might find it possible to split the larger square into two. After that it would be easy to arrange all the shapes into an overall simple shape. This analogy is only meant to illustrate how sometimes a problem cannot be solved by trying different arrangements of the given pieces but only by reexamination of the pieces themselves.

If the above problem was actually set as a problem and the solution given as indicated there would immediately be an outcry that this was 'cheating'. There would be protests that it was assumed that the given shapes could not themselves be altered. Such a cry of cheating always

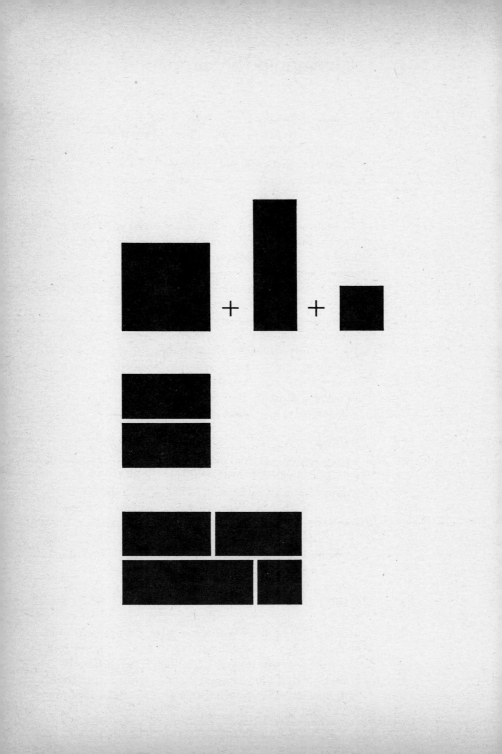

reveals the use of certain assumed boundaries or limits.

In problem solving one always assumes certain boundaries. Such boundaries make it much easier to solve the problem by reducing the area within which the problem solving has to take place. If someone were to give you an address in London it might be hard to find. If someone told you it was north of the Thames it would be slightly easier to find. If someone told you that it was within walking distance of Piccadilly Circus it would be that much easier to find. So it is with problem solving that one sets one's own limits within which to explore. If someone else comes along and solves the problem by stepping outside the limits there is an immediate cry of 'cheating'. And yet the limits are usually self-imposed. Moreover they are imposed on no stronger grounds than that of convenience. If such boundaries or limits are wrongly set then it may be as impossible to solve the problem as it would be to find an address south of the river Thames by looking north of the river.

Since it would be quite impossible to reexamine everything in sight one has to take most things for granted in any situation – whether or not it is a problem situation. Late one Saturday morning I was walking down a shopping street when I saw a flower seller holding out a large bunch of carnations for which he was only charging two shillings (ten newpence). It seemed a good bargain and I assumed that it was the end of the morning and he was getting rid of his leftover flowers. I paid him whereupon he detached a small bunch of about four carnations from the large bunch and handed them to me. The little bunch was a genuine bunch wrapped with a little bit of wire. It was only my greed that had assumed that the bunch offered had referred to the whole bunch he held in his hand.

A new housing estate had just been completed. At the ceremonial opening it was noticed that everything appeared to be a little bit low. The ceilings were low, the doors were low, the windows were low. No one could understand what had happened. Finally it was discovered that someone had sabotaged the measuring sticks used by the workmen by cutting an inch off the end of each one. Naturally everyone using the sticks had assumed that they at least were correct since they were used to show the correctness of everything else.

There is made in Switzerland a pear brandy in which a whole pear is to be seen within the bottle. How did the pear get into the bottle? The usual guess is that the bottle neck has been closed after the pear has been put into the bottle. Others guess that the bottom of the bottle was added after the pear was inside. It is always assumed that since the pear is a fully grown pear that it must have been placed in the bottle as a fully grown pear. In fact if a branch bearing a tiny bud was inserted through the neck of the bottle then the pear would actually grow within the bottle and there would be no question of how it got inside.

In challenging assumptions one challenges the necessity of boundaries and limits and one challenges the validity of individual concepts. As in lateral thinking in general there is no question of attacking the assumptions as wrong. Nor is there any question of offering better alternatives. It is simply a matter of trying to restructure patterns. And by definition assumptions are patterns which usually escape the restructuring process.

Practice session
1   Demonstration problems
Problem
A landscape gardener is given instructions to plant four special trees so that each one is exactly the same distance from each of the others. How would you arrange the trees?

The usual procedure is to try and arrange four dots on a piece of paper so that each dot is equidistant from every other dot. This turns out to be impossible. The problem seems impossible to solve.

The assumption is that the trees are all planted on a level piece of ground. If one challenges this assumption one finds that the trees can indeed be planted in the manner specified. But one tree is planted at the top of a hill and the other three are planted on the sides of the hill. This makes them all equidistant from one another (in fact they are at the angles of a tetrahedron). One can also solve the problem by placing one tree at the bottom of a hole and the others around the edge of the hole.

Problem
This is an old problem but it makes the point very nicely. Nine dots are arranged as shown overleaf. The problem is to link up these nine dots using only four straight lines which must follow on without raising the pencil from the paper.

At first it seems easy and various attempts are made to link up the dots. Then it is found that one always needs more than four. The problem seems impossible.

The assumption here is that the straight lines must link up the dots and must not extend beyond the boundaries set by the outer line of dots. If one breaks through this assumption and does go beyond the boundary then the problem is easily solved as shown.

Problem
A man worked in a tall office building. Each morning he got in the lift on the ground floor, pressed the lift button to the tenth floor, got out of the lift and walked up to the fifteenth floor. At night he would get into the lift on the fifteenth floor and get out again on the ground floor. What was the man up to?

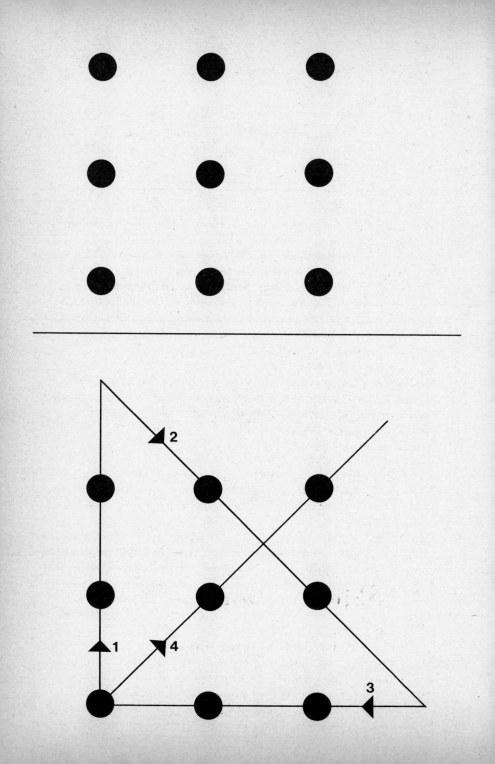

Various explanations are offered. They include:
The man wanted exercise.
He wanted to talk to someone on the way up from the
tenth to the fifteenth floor.
He wanted to admire the view as he walked up.
He wanted people to think he worked on the tenth floor
(it might have been more prestigious) etc.
In fact the man acted in this peculiar way because he
had no choice. He was a dwarf and could not reach
higher than the tenth floor button.

The natural assumption is that the man is perfectly
normal and it is his behaviour that is abnormal.

One can generate other problems of this sort. One can
also collect examples of behaviour which seem bizarre
until one knows the real reason behind it. The purpose
of these problems is just to show that the acceptance of
assumptions may make it difficult or impossible to solve
a problem.

2   The block problems
Problem
Take four blocks (these may be matchboxes, books,
cereal or detergent packets). The problem is to arrange
them in certain specified ways. These ways are specified
by how the blocks come to touch each other in the
arrangement. For two blocks to be regarded as touching
any part of any flat surface must be in contact—a corner
or an edge does not count.
The specified arrangements are as follows:
1   Arrange the blocks so that each block is touching
two others.
2   Arrange the blocks so that one block is touching one
other, one block is touching two others, and another
block is touching three others.
3   Arrange the blocks so that each block is touching
three others.

4   Arrange the blocks so that each block is touching one other.

Solutions

1   There are several ways of doing this. One way is shown opposite. This is a 'circular' arrangement in which each block has two touching neighbours – one in front and one behind.

2   There is often some difficulty with this one because it is *assumed* that the problem has to be solved in the sequence in which it was posed i.e. one block to touch one other, one block to touch two others, one block to touch three others. If however, a start is made at the other end by making one block touch three others then this arrangement can be progressively modified to give the arrangement shown.

3   Some people have a lot of difficulty with this problem because they *assume* that all the blocks have to lie in the same plane (i.e. spread out on the surface being used). As soon as one breaks free of this assumption and starts to place the blocks on top of one another one can reach the required arrangement.

4   There is a surprising amount of difficulty in solving this problem. The usual mistake is to arrange the blocks in a long row. In such a row the end blocks are indeed touching only one other but the middle blocks have two neighbours. A few people actually declare that the problem can not be solved. The correct arrangement is very simple.

Comment

Most people solve the block arranging problems by playing around with the blocks and seeing what turns up. Nothing much would happen if one did this without bothering to have the blocks touching one another. So for convenience one assumes that the blocks all have to touch one another in some fashion (i.e. there has to be a single arrangement). It is this artificial limit, this assumption, that makes it so difficult to solve the last problem which is so easy in itself.

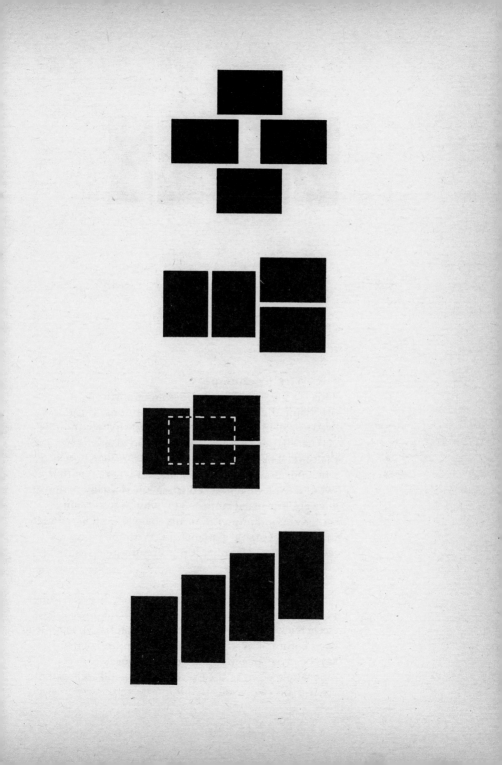

The 'Why' Technique

This is a game which provides an opportunity for practising the challenging of assumptions. It can also be used as a deliberate technique. The 'why' technique is very similar to the usual child's habit of asking 'why' all the time. The difference is that 'why' is usually asked when one does not know the answer whereas with the 'why' technique it is asked when one does know the answer. The usual response to 'why' is to explain something unfamiliar in terms that are familiar enough to be an acceptable explanation. With the 'why' technique these familiar terms are questions as well. Nothing is sacred.

The process is rather more difficult than it seems. There is a natural tendency to run out of explanations or to circle back and give an explanation that has already been used before. There is also the very natural tendency to say 'because' if something very obvious is questioned. The whole point of the exercise is to avoid feeling that anything is so obvious that it merits a 'because' answer.

The teacher makes some sort of statement and then a student asks, 'Why?'. The teacher offers an explanation which is in turn met by another, 'Why?'. If the process was no more than an automatic repetition of 'why' then one would hardly need a second party to ask why except that the student gets into the habit of assuming nothing. In practice it never is an automatic repetition of 'why'. The question is directed to some particular aspect of the previous explanation rather than being a blanket response. 'Why' can be focused.

Examples
Why are blackboards black?
Because otherwise they would not be called blackboards.
Why would it matter what they were called?
It would not matter.
Why?
Because they are there to write or draw upon.
Why?
Because if something is to be shown to the whole classroom it is easier to write on the blackboard where everyone can see it.

The above questioning might however have taken quite a different line.
Why are blackboards black?
So that the white chalk marks can be seen easily.
Why do you want to see the white chalk marks?
or:
Why is the chalk white?
or:
Why does one want to use white chalk?
or:
Why don't you use black chalk?

In each of these cases 'why' is directed to a particular aspect of the subject and this determines the development of the questioning. The teacher can of course also direct the development by the way the question is answered.

The teacher keeps up the answers as long as possible. He may however at any time say: 'I don't know. Why do you think?' If the student can give an answer then the roles can be reversed with the student answering the why questions and the teacher putting them.

Some possible subjects for this type of session are given below:
Why are wheels round?
Why does a chair have four legs?
Why are most rooms square or oblong?
Why do girls wear different clothes from boys?
Why do we come to school?
Why do people have two legs?

The usual purpose of 'why' is to elicit information. One wants to be comforted with some explanation which one can accept and be satisfied with. The lateral use of why is quite opposite. The intention is to create discomfort with any explanation. By refusing to be comforted with an explanation one tries to look at things in a different way and so increases the possibility of restructuring the pattern.

In answering the question the teacher does not have to struggle to justify something as a unique explanation. In his answer he can suggest alternatives. The answer to the question, 'Why does the blackboard have to be black?' could be, 'It does not have to be black, it could be green or blue so long as the white chalk showed.' The impression that there is a unique and necessary reason behind everything must be avoided. Contrast the answers:
'Blackboards are black because black is a convenient colour to show up white chalk marks.'

'Blackboards are black because otherwise you would not see what was written on them.'

Even if there is a true historic reason behind something the teacher must not give the impression that the historic reason is a sufficient one. Suppose that blackboards really were black because the usefulness of white chalk was discovered first. Historically this is an accurate reason for the use of black but in practice it is not enough. After all it only explains why people started to use black but does not explain why it is convenient to continue doing so. One might say: 'Blackboards were originally coloured black because they were looking for a surface to show up the white chalk marks. They have continued to be black ever since because black has proved satisfactory.'

Summary

In dealing with situations or problems many things have to be taken for granted. In order to live at all one must be making assumptions all the time. Yet each of these assumptions is a cliché pattern which may be restructured to make better use of available information. In addition the restructuring of more complex patterns may prove impossible unless one breaks through some assumed boundary. The idea is to show that any assumption whatsoever can be challenged. It is not a matter of pretending that one has time to challenge every assumption on every occasion but of showing that nothing is sacred.

The idea is not to sow so much doubt that one is reduced to dithering indecision through being unable to take anything for granted. On the contrary one acknowledges the great usefulness of assumptions and clichés. In fact one is much freer to use assumptions and clichés if one knows that one is not going to be imprisoned by them.

Innovation

The two preceding chapters have been concerned with
two fundamental aspects of the lateral thinking process:
● The deliberate generation of alternative ways of
  looking at things.
● The challenging of assumptions.
  In themselves these processes are not far removed from
  ordinary vertical thinking. What is different is the
  'unreasonable' way in which the processes are applied
  and the purpose behind the application. Lateral
  thinking is concerned not with development but with
  restructuring.

Both the processes mentioned above have been applied
for the purpose of description or analysis of a situation.
This could be called backward thinking: this is a matter
of looking at something that is there and working it over.
Forward thinking involves moving forward. Forward
thinking involves building up something new rather
than analysing something old. Innovation and creativity
involve forward thinking. The distinction between
backward and forward thinking is entirely arbitrary.
There is no real distinction because one may have to
look backward in a new way in order to move forward.
A creative description may be just as generative as a
creative idea. Both backward thinking and forward
thinking are concerned with alteration, with
improvement, with bringing about some effect. In
practice backward thinking is however more concerned
with explaining an effect whereas forward thinking is
more concerned with bringing about an effect.

Before going on to consider innovation it is necessary to
consider an aspect of thinking that applies much more
to forward thinking than to backward thinking. This is
the matter of evaluation and suspended judgment.

The purpose of thinking is not to be right but to be effective. Being effective does eventually involve being right but there is a very important difference between the two. Being right means being right all the time. Being effective means being right only at the end.

Vertical thinking involves being right all along. Judgment is exercised at every stage. One is not allowed to take a step that is not right. One is not allowed to accept an arrangement of information that is not right. Vertical thinking is selection by exclusion. Judgment is the method of exclusion and the negative ('no', 'not') is the tool of exclusion.

With lateral thinking one is allowed to be wrong on the way even though one must be right in the end. With lateral thinking one is allowed to use arrangements of information which are invalid in themselves in order to bring about a restructuring that is valid. One may have to move to an untenable position in order to be able to find a tenable position.

In lateral thinking one is not so concerned with the nature of an arrangement of information but with where it can lead one. So instead of judging each arrangement and allowing only those that are valid one suspends judgment until later on. It is not a matter of doing without judgment but of deferring it until later.

As a process lateral thinking is concerned with change not with proof. The emphasis is shifted from the validity of a particular pattern to the usefulness of that pattern in generating new patterns.
There is nothing 'unreasonable' about the other lateral thinking processes described so far but the need to suspend judgment is so fundamentally different from vertical thinking that it is much harder to understand.

Education is soundly based on the *need to be right all
the time*. Throughout education one is taught the
correct facts, the correct deductions to be made from
them and the correct way of making these deductions.
One learns to be correct by being made very sensitive to
what is incorrect. One learns to apply judgment at
every stage and to follow up this judgment with the
'no' label. One learns how to say, 'no', 'this is not so',
'this cannot be so', 'this does not lead to that', 'you are
wrong here', 'this would never work', 'there is no reason
for that' and so on. This sort of thing is the very essence
of vertical thinking and accounts for its great usefulness.
The danger lies in the arrogance of the attitude that
assumes that vertical thinking is sufficient. It is not.
Exclusive emphasis on the need to be right all the time
completely shuts out creativity and progress.

The need to be right all the time is the biggest bar there
is to new ideas. It is better to have enough ideas for
some of them to be wrong than to be always right by
having no ideas at all.

The need to make use of provocative arrangements of
information in order to bring about insight repatterning
is dictated by the behaviour of mind as a
self-maximizing memory system*. In practice this need
is met by delaying judgment. Judgment is suspended
during the generative stage of thinking in order to be
applied during the selective stage. The nature of the
system is such that a wrong idea at some stage can lead
to a right one later on. Lee de Forest discovered the
immensely useful thermionic valve through following
up the erroneous idea that an electric spark altered the
behaviour of a gas jet. Marconi succeeded in
transmitting wireless waves across the Atlantic ocean
through following up the erroneous idea that the waves
would follow the curvature of the earth.

The major dangers of the need to be right all the time arc as follows:

- Arrogant certainty attends a line of thought which though correct in itself may have started from wrong premises.
- An incorrect idea which would have led on to a correct idea (or useful experimentation) is choked off at too early a stage if it cannot itself be justified.
- It is assumed that being right is enough – an *adequate* arrangement blocks the possibility of a better arrangement.
- The importance attached to being right all the time breeds the inhibiting fear of making mistakes.

Delay in judgment
A later chapter deals with the lateral process which involves being wrong on purpose in order to provoke a rearrangement of information. What is being considered here is simply the *delaying of judgment* instead of applying it immediately. In practice judgment may be applied at any of the following stages:

- Judgment as to whether an information area is relevant to the matter under consideration. This precedes the development of any ideas.
- Judgment as to the validity of an idea in one's own internal thinking process. Dismissing such an idea instead of exploring it.
- Judgment as to its correctness before offering an idea to others.
- Judgment of an idea offered by someone else – either in refusing to accept it or in actual condemnation of it.

In this regard judgment, evaluation and criticism are regarded as similar processes. Suspension of judgment does not imply suspension of condemnation – it implies suspension of judgment whether the outcome is favourable or otherwise.

The suspension of judgment can have the following effects:

● An idea will survive longer and will breed further ideas.
● Other people will offer ideas which their own judgment would have rejected. Such ideas may be extremely useful to those receiving them.
● The ideas of others can be accepted for their stimulating effect instead of being rejected.
● Ideas which are judged to be wrong within the current frame of reference may survive long enough to show that the frame of reference needs altering.

In the diagram opposite A is the starting point of a problem. In tackling the problem one moves towards K but this idea is unsound and so it is rejected. Instead one moves towards C. But from C one can go nowhere. Had one moved towards K then one could have proceeded from there to G and from G to B which is the solution. Once one had reached B then one would have been able to see the correct path from A through P.

Practical application

The principle of suspended judgment has been discussed. The practical application of this principle needs outlining for it is not much use accepting the principle but never applying it. In practice the principle leads to the following behaviour:

● One does not rush to judge or evaluate an idea. One does not regard judgment or evaluation as the most important thing that can be done to an idea. One prefers exploration.
● Some ideas are obviously wrong even when no attempt at judgment is made. In such cases one shifts attention from why it is wrong to how it can be useful.
● Even if one knows that an idea must eventually be thrown out one delays that moment in order to extract

as much usefulness from the idea as possible.
● Instead of forcing an idea in the direction which
judgment indicates one follows along behind it.

A bucket with holes cannot carry much water. One
could reject it out of hand. Or one could see how far it
could carry how much water. In spite of the holes it may
be very useful for bringing about a certain effect.

In so far as it is not just a matter of copying, design requires a good deal of innovation. Design is a convenient format for practising the lateral thinking principles that have been discussed up to this point. The design process itself is discussed at length in a later section; in this section design is used as practice for lateral thinking.

Practice

The designs are to be visual and in black and white or colour. Verbal descriptions can be added to the pictures to explain certain features or to explain how they work. The advantages of a visual format are many.

1   There has to be a definite commitment to a way of doing something rather than a vague generalized description.
2   The design is expressed in a manner that is visible to everyone.
3   Visual expression of a complicated structure is much easier than verbal expression. It would be a pity to limit design by the ability to describe it.

The designs could be worked out as a classroom exercise or they could be done as homework. It is easier if the students all work on the same design rather than on individual choices for then any comments apply to them all, there is more comparison and they are all more involved in the analysis.

It is convenient if all the designs are executed on standard sized sheets of paper. Once the design task has been set no additional information is given. No attempt is made to make the design project more specific. 'Do whatever you think is best' is the answer to any question.

● Comment on results
Unless the group is small enough to actually cluster

around the drawings these would have to be copied and
shown on an overhead projector or epidiascope. Or they
could just be pinned up. Adequate discussion could be
carried out without showing the drawings at all but just
redrawing the important features on the blackboard. In
commenting on the results the teacher would want to
bear the following points in mind:
1   Resist the temptation to judge. Resist the temptation
to say, 'this would not work because . . . .'
2   Resist the temptation to choose one way of doing
things as being much better than any other for fear of
polarizing design in one direction.
3   Emphasize the variety of the different ways of
carrying out a particular function. List the different
suggestions and add others of one's own.
4   Try and look at the function underlying a particular
design. Try to separate the intention of the designer
from the actual way this was carried out.
5   Note the features that have been put there for a
functional purpose and the ones that are there as
ornaments to complete the picture.
6   Question certain points – not in order to destroy
them but in order to find out if there was any special
reason behind them which may not be manifest.
7   Note the borrowing of complete designs from what
might have been seen on television, in the cinema or in
comics.

● Suggestions
Design projects can either ask for improvements on
existing things or for the actual invention of something
to carry out a task. It is easiest if the designs do involve
something physical since this is easier to draw. They do
not have to be mechanical in the strictest sense of the
word, for instance the design of a new classroom or a
new type of shoe would be very suitable. It is enough
that they are concrete projects. In addition one can try
organizational designs. Organizational designs would

ask for ways of doing things such as building a house very quickly.

Design:
An apple picking machine.
A potato peeling machine.
A cart to go over rough ground.
A cup that cannot spill.
A machine to dig tunnels.
A device to help cars to park.

Redesign:
The human body.
A new milk bottle.
A chair.
A school.
A new type of clothes.
A better umbrella.

Organizational:
How to build a house very quickly.
How to arrange the checkout counters in a supermarket.
How to organize garbage collection.
How to organize shopping to take up the least time.
How to put a drain across a busy road.

● Variety
The purpose of the design session is to show that there can be different ways of doing something. It is not the individual designs that matter so much as the comparison between designs. In order to show this variety one could compare the complete designs but it is more effective to pick out some particular function and show how this was handled by the different designers. For instance in the design of an apple picking machine one could choose the function of 'reaching the apples'. To reach the apples some students will have used extendable arms, others will have raised the whole

vehicle on jacks, others will have tried to bring the
apples to the ground, others might have planted the
trees in trenches anyway. For each function the teacher
lists the different methods used and asks for further
suggestions. He can also add suggestions of his own or
ones derived from previous experience with the design
project.

Particular functions with the apple picking machine
could include the following:
Reaching the apples.
Finding the apples.
Picking the apples.
Transporting the apples to the ground.
Sorting out the apples.
Putting the apples in containers.
Moving onto the next tree.

It is not suggested that in carrying out the design the
student will have tried to cover all these functions. Most
of them would be covered quite unconsciously.
Nevertheless one can consciously analyse what has been
done and show the different ways of doing it. In many
cases no provision will have been made for carrying out
a certain function (e.g. transporting the apples to the
ground). In such cases one does not criticize the designs
that do not show the function but commends those that
do show it.

● Evaluation
One could criticize designs for omissions, for errors of
mechanics, for errors of efficiency, for errors of
magnitude and for all sorts of other errors. It is difficult
to resist the temptation to do this—but the temptation
must be resisted.

If some designs have left things out then one shows this
up by commenting on those designs which have put it in.

If some design shows an arrangement that is mechanically unsound then one comments on the function intended rather than the particular way of carrying it out.

If some designs show a very roundabout way of doing something one describes the design without criticism and then describes more efficient designs.

One of the most common faults with designs by students in the 10–13 age group is the tendency to lose sight of the design project and to go into great detail drawing some vehicle that is derived directly from another source such as television or space comics. Thus an apple picking machine will be shown bristling with guns, rockets, radar and jets. Details will be given about number of crew, speed, range, power, how much it would cost to build, how long it would take to build, how many nuts and bolts, the materials used in construction and so on. There is no point in criticizing the superfluity of all this. Instead one emphasizes the functional economy and effectiveness of other designs.

It is important not to criticize actual mechanics. One designer of an apple picking machine suggested putting bits of metal in each of the apples and then using powerful magnets buried in the ground under each tree to pull the apples down. It would be easy to criticize this as follows:

1   Just as much trouble to put bits of metal in each apple as to pick each one directly.
2   The magnet would have to be very powerful indeed to pull the apples down from such a distance.
3   The apples would be badly damaged on hitting the ground.
4   Buried magnets would only be able to collect apples from one tree.

These are all valid comments and one could make many
more. But rather than criticizing in this manner one
could say: 'Here is someone who instead of going up to
pick the apples like everybody else wants to attract the
apples to the ground. Instead of having to find the
apples and then to pick them one by one he can get them
all together and all at once.' Both these are very valid
points. The actual method for carrying out the function
is obviously inefficient but it is better to let that be than
to appear to criticize the concept of function by
criticizing the way it is carried out. When that particular
designer learns more about magnets he will find that
they would not be much good. At the moment however
they represent the only method he knows for carrying
out 'attraction from a distance'.

In another design for a cart that would go over rough
ground the designer suggested some sort of 'smooth
stuff' that was sucked up by the cart from behind itself
and then spread down in front of it. Thus the cart was
always travelling over smooth stuff. There was even a
reservoir for evening out the supply of the smooth stuff.
It would be easy to criticize the idea as follows:
1   What sort of 'smooth stuff' would fill in big hollows.
One would need far too much.
2   One could never suck back all that had been laid
down and so the supply would run out after a few feet.
3   The cart would have to move very slowly indeed.

Such criticisms are easy but instead one would
appreciate that the designer had got away from the
usual approach of providing special wheels or other
devices for going over rough ground and instead was
trying to alter the *ground itself*. From such a concept
could come the notion of a tracked vehicle which does
actually lay down smooth stuff and pick it up again.
There are also those military vehicles which have a roll
of steel mesh or glass fibre matting on their backs and

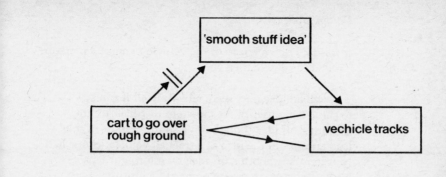

Paul
tant.

A machine for travelling over rough ground.

hand goes on
which steers a wheel

A Screen
where
he can
See
where
he is going!

Sucks
Smooth
Suff up
which goes
down to
Kind of slide to where
it is Stored

foot goes on a
pedal to Start up.

Smooth
Suff is Stored here

lays Smooth
Suff down

this is laid down ahead of the vehicle to make a road on which the vehicle then runs.

Though an idea may seem silly in itself it can still lead to something useful. As shown in the diagram the smooth stuff idea though not a solution in itself might lead straight to the idea of a tracked vehicle. If one had rejected the smooth stuff idea then it might have been harder to get to the same point. The attitude is not, 'This won't work let's throw it out' but, 'This is not going to work but what does it lead us to.'

No one is silly for the sake of being silly no matter how it might appear to other people. There must be a reason why something made sense to the person who drew it at the moment when it was drawn. What it appears to other people is not so important if one is trying to encourage lateral thinking. In any case whatever the reason behind a design and however silly it may be it can still be a most useful stimulus to further ideas.

● Assumptions
In the design process there is a tendency to use 'complete units'. This means that when one borrows a unit from somewhere else in order to carry out some special function that unit is used 'complete'. Thus a mechanical arm to pick apples will have five fingers because the human arm has that number. In an attempt to break up such complete units and isolate what is really required one can question the assumption behind them: 'Why does a hand need five fingers to pick apples?

One may also question assumptions that seem to be basic to the design itself.
Why do we have to *pick* the apples off the trees?
Why do trees have to be that shape?
Why does the arm have to go up and down with every apple it picks?

Some of the points challenged could easily have been taken for granted. By challenging them one can open up new ideas. For instance one could shake apples from trees instead of picking them. In California, they are experimenting with growing trees in a special way which would make it possible to pick the fruit more easily. The arm does not need to go up and down with each apple, the apples could be dropped into a chute or container.

The 'why' technique can be applied to any part of the design project. To begin with the teacher would apply it after discussing the designs. The students could also apply it to their own designs or those of others. As usual the purpose of the 'why' technique is not to try and justify something but to see what happens when one challenges the uniqueness of a particular way of doing things.

Summary
The design process is a convenient format for developing the idea of lateral thinking. The emphasis is on the *different* ways of doing things, the *different* ways of looking at things and the escape from cliché concepts, the challenging of assumptions. Critical evaluation is temporarily suspended in order to develop a generative frame of mind in which flexibility and variety can be used with confidence. For the design session to work it is essential that the person running it understands the purpose of the session. It is not practice in design but practice in lateral thinking.

Dominant ideas
and crucial factors

12

There is nothing vague about a geometrical shape. As a
situation it is very definite – one knows what one is
looking at. Most situations however are much more
vague than this. Most of the time one has a vague
awareness of the situation and nothing more. With a
definite geometrical shape it is easy to think of
alternative ways of dividing it up and alternative ways of
putting the pieces together again. It is much more
difficult to do this if there is only a vague awareness of
the situation.

Everyone is confident that they know what they are
talking about, reading about or writing about but if you
ask them to pick out the dominant idea there is
difficulty in doing so. It is difficult to convert a vague
awareness into a definite statement. The statement is
either too long and complicated or else it leaves out too
much. Sometimes the different aspects of the subject do
not hang together to give a single theme.

Unless one can convert a vague awareness to a definite
pattern it is extremely difficult to generate alternative
patterns, alternative ways of looking at the situation. In
a defining situation one picks out the dominant idea
not in order to be frozen by that idea but in order to be
able to generate alternative ideas.

Unless one can pick out the dominant idea one is going
to be dominated by it. Whatever way one tries to look at
the situation is likely to be dominated by the ever
present but undefined dominant idea. One of the main
purposes of picking out the dominant idea is to be able
to escape from it. One can more easily escape from
something definite than from something vague.
Liberation from rigid patterns and the generation of
alternative patterns are the aims of lateral thinking.
Both processes are made much easier if one can pick out
the dominant idea.

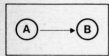

If one can not pick out the dominant idea then any alternatives one generates are likely to be imprisoned within that vague general idea. The diagram shows how one may feel that one is generating an alternative point of view and yet this is still within the same framework of the dominant idea as the original point of view. It is only when one becomes aware of the framework that one can generate an alternative point of view outside of it.

The dominant idea resides not in the situation itself but in the way it is looked at. Some people seem much better at picking out the dominant idea. Some people seem much better at crystallizing the situation in a single sentence. This may be because they can separate the main idea from the detail or it may be because they tend to have a simpler view of things. In order to be able to pick out the dominant idea one must make a conscious effort to do so and one needs practice.

Different dominant ideas
If students are asked to pick out the dominant idea from a newspaper article there are usually several different versions of what the dominant idea is. From an article on parks the following may be chosen as the dominant idea:
The beauty of parkland.
The value of parkland as a contrast to the city surroundings.
The need to develop more parks.
The difficulty of developing or preserving parks.
Parkland as a relaxation or pleasure.
The author is exercising the function of protest and parkland happens to be a suitable theme.
The danger of the demands of urban growth.

These are all different but related ideas. It is easy to say that some of the ideas are more truly dominant than others and yet to the person picking out the idea this idea has a valid dominance. It is not a matter of finding *the* dominant idea but of getting into the habit of trying to pick out the dominant idea. It is not a matter of analysing the situation but of seeing it clearly enough to be able to generate different points of view. It is not a matter of making use of the dominant idea but of identifying it in order to avoid it.

In the design situation discussed in the preceding section the organizing effect of the dominant idea is quite obvious. The dominant idea is never actually stated but for different groups the idea is different. When children try to design a machine for picking apples the dominant idea is 'reaching the apples'. The children think in personal terms which involve wanting one apple at a time and also the difficulty (for a small child) of actually reaching the apples. When the same design problem is given to an industrial engineering group the dominant idea is 'effectiveness in commercial terms'. This is a wide concept which includes speed and cheapness of operation without any damage to the apples. From this point of view reaching the apples is not so much a problem as finding them, picking many at a time, bringing them to the ground without damage, and all this with a cheap machine that can easily be moved from tree to tree. In short the dominant problem for the engineers is 'advantage over manual labour' whereas for the children it was 'getting the apples'.

Hierarchy of dominant ideas

As soon as one starts picking out dominant ideas one becomes aware that there are different degrees of comprehensiveness of dominant ideas. The dominant idea may include the whole subject or only one aspect of it. Thus from an article on crime one might pick out the following dominant ideas:
Crime.
Behaviour of people.

Violence.
Social structures and crime.
The trend of crime.
What can be done.

Clearly 'crime' and 'the behaviour of people' are much wider ideas than 'violence' or 'what can be done' but all of them are valid dominant ideas. There is a hierarchy which extends upwards from the more specific ideas to the more general. In picking out the dominant idea it is not a matter of searching for the most general and most comprehensive idea for this may be so very wide that it is impossible to move outside it at all. In picking out a dominant idea it is not a matter of having to justify to someone else that the idea is *the dominant idea* which covers the whole situation and which can not therefore be challenged. It is a matter of picking out an idea which seems (to oneself) to dominate the issue. For instance in the article on crime the dominant idea might have seemed to be 'the uncertainty about the value of punishment' or 'the protection of the rights of a citizen even if he was a criminal'.

Crucial factor

dominant idea

A dominant idea is the organizing theme in a way of looking at a situation. It is often present but undefined and one tries to define it in order to escape from it. A crucial factor is some element of the situation which must always be included no matter how one looks at the situation. The crucial factor is a tethering point. Like a dominant idea a crucial factor can immobilize a situation and make it impossible to shift a point of view. Like a dominant idea a crucial factor may exert a powerful influence without ever being consciously recognized.

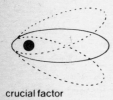

crucial factor

The difference between a dominant idea and a crucial factor is shown diagrammatically opposite. The dominant idea organizes the situation. The crucial factor tethers it and though some mobility is allowed this is restricted.

The purpose of isolating crucial factors is to examine them. Very often a crucial factor is an assumption – at least the 'crucial' nature of that factor is an assumption. Once the factor is isolated one challenges the necessity for it. If the factor is found not to be crucial then the tethering effect of that factor disappears and there is more freedom in structuring the situation in a different way. In the design of a machine for picking apples a crucial factor may have been 'that the apples must not be damaged' or 'that only ripe apples were to be picked'. The necessity to include such crucial factors would restrict the way the problem could be looked at. For instance shaking the tree would not be a good idea.

There may be one crucial factor, several crucial factors or none at all. Different people may choose different crucial factors. As with finding the dominant idea what matters is that one identifies what seems to be a crucial factor in one's own view of the problem. Whether it really is crucial or whether other people would think so does not matter for one picks it out only to challenge its necessity.

In looking for the dominant idea one wants to know, 'why are we always looking at this thing in the same way'. In looking for the crucial factor one wants to know, 'what is holding us up, what is keeping us to this old approach?'

In itself the search for dominant ideas or crucial factors is not a lateral thinking process at all. It is a necessary step which allows one to use lateral thinking more effectively. It is difficult to restructure a pattern unless one can see the pattern. It is difficult to loosen up a pattern unless one can identify the rigid points.

Practice

1   A newspaper article is read out to the students who
then have to note down:
(1)   The dominant idea (or ideas).
(2)   The crucial factors.

When the results have been collected the teacher goes
through them and lists the different choices. A person
making a particular choice may be asked to explain why
he made that choice. This is not in order that the choice
be justified or in order to show that it was not as good as
other choices but in order to elaborate a particular point
of view. There is no attempt to disqualify any of the
choices or to rank them in order of excellence.

If it is clear that some of the students have not grasped
the point about dominant ideas and crucial factors then
one concentrates on those answers which make the
point most clearly. If none of them do then the teacher
has to supply his own choice of dominant idea and
crucial factor for the passage used.

It is not a good idea to ask for choices of dominant ideas
and then to list them on the blackboard as was suggested
in previous sections. This is because a choice which
seems to be very good will inhibit any further
suggestions. It is far better to let people work out what
seems dominant or crucial to them and then to show the
variety of answers.

2   Radio or tape recorder
Instead of the teacher reading the passage out it could
be a feature programme on the radio or something
taped off the radio. The advantage of a tape recording is
its repeatability.

3   Instead of listening to a passage being read out the students can be given passages to study for themselves. This is rather different since there is more time to go over the piece, the interpretation is not so determined by the way it has been read out and one can go back and reexamine what has been written to see if it supports a particular point of view.

4   Discussion
Two students are asked to debate a subject in front of the class. One can either choose students who declare they have opposite views on a particular subject or else ask the students to debate from opposite points of view whether or not they hold those views. The rest of the class listens to the debate and notes down the dominant idea and crucial factors in the discussion. In order to try and check the validity of these the other students can ask the debaters questions.

5   Design project
Either in the course of a design project or in discussing the results of a design project undertaken by others the students can try and pick out the dominant ideas and crucial factors. In this case they can examine the crucial factors to see whether they really are crucial and to see what would happen if one did not include them in the design. The same thing could be done with dominant ideas: the students first picking out the ideas and then seeing how they could escape them.

Although it would be easy to combine this sort of practice session with the lateral thinking processes described before (and those to be described later) it is probably better not to do so. If one were to combine the process of generating alternatives with the process of picking out the dominant idea then there is a tendency to pick out a dominant idea which fits nicely in with the alternative one can think of. The choice of dominant ideas and crucial factors soon becomes tailored to show how clever one is at avoiding them. For the moment it is enough to become skilled at finding dominant ideas and crucial factors.

Fractionation

13

The aim of lateral thinking is to look at things in different ways, to restructure patterns, to generate alternatives. The mere intention of generating alternatives is sometimes sufficient. Such an intention can make one pause and look around before proceeding too far with the obvious way of looking at the situation. As one looks around one may find that there are other alternatives waiting to be considered. At other times the mere intention of generating alternatives is not sufficient. Goodwill cannot by itself generate alternatives. One has to use some more practical method. For the same reason exhorting people to look for alternatives does have a certain usefulness (especially in tempering the arrogance of a unique point of view) but one also needs to develop ways of generating alternatives.

In the self-maximizing memory system of mind there is a tendency for established patterns to grow larger and larger. The patterns may grow by extension or else two separate patterns may join up to form a large single one. This tendency of patterns to grow larger is seen clearly with language. Words describing individual features are put together to describe a new situation which soon acquires its own language label. Once this has happened a new standard pattern has been formed. This new pattern is used in its own right without constant reference to the original features which made up the pattern.

The more unified a pattern the more difficult it is to restructure it. Thus when a single standard pattern takes over from a collection of smaller patterns the situation becomes much more difficult to look at in a new way. In order to make such restructuring easier one tries to return to the collection of smaller patterns. If a child is given a complete doll's house he has little choice but to use and admire it as it is. If however he is given a box of building blocks then he can assemble them in different ways to give a variety of houses.

Opposite is shown a geometrical shape which could be described as an 'L shape'. The problem is to divide this shape into four pieces which are exactly similar in size, shape and area. Initial attempts to do this usually take the form of the divisions shown. These are obviously inadequate since the pieces are not the same in size even though they may be the same in shape.

A correct solution is shown overleaf and it is seen to consist of four small L shaped pieces. An easy way to reach this answer is to divide the original shape into three squares and then to divide each of these into four pieces which gives a total of twelve pieces. These twelve pieces must then be assembled in four groups of three and when this has been accomplished as shown, the original is divided into the required four pieces.

One of the problems set in a previous chapter asked for a square to be divided into four pieces which were the same in size, shape and area. Some people went further than the usual obvious divisions by dividing the square into sixteen small squares and then reassembling them in different ways to give a variety of new ways of dividing the square into four.

In a sense the whole point of language is to give separate units that can be moved around and put together in different ways. The danger is that these different ways soon become established as fixed units themselves and not as temporary arrangements of other units.

If one takes any situation and breaks it down into fractions one can then restructure the situation by putting the fractions together in a new way.

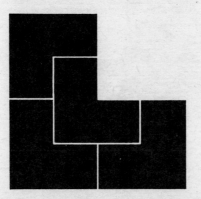

True and false divisions

It might seem that what is being recommended is the analysis of a situation into its component parts. This is not so. One is not trying to *find* the true component parts of a situation, one is trying to *create* parts. The natural or true lines of division are usually not much good as the parts tend to reassemble to give the original pattern since this is how the pattern came about in the first place. With artificial divisions however there is more opportunity to put units together in novel ways. As is so often the case with lateral thinking one is looking for a provocative arrangement of information that can lead to a new way of looking at things. One is not trying to discover the *correct* way. What one needs is something to be going on with and for this purpose any sort of fractionation will do.

In the design of an apple picking machine the problem could have been fractionated into the following parts:
reaching
finding
picking
transport to the ground
undamaged apples.

In reassembling these fractions one might have put reaching-finding-picking together and then substituted shaking the tree for all these functions. One would then be left with transport to the ground in such a way that the apples were not damaged. On the other hand one might have put reaching-undamaged apples-transport to the ground together and come up with some elevated canvas platform which would be raised towards the apples.

Someone else might have fractionated the problem in a different way:
contribution of tree to apple picking

contribution of apples
contribution of machine
This particular type of fractionation might have led on
to the idea of growing the trees in a special way that
would make it easier to pick the apples.

Complete division and overlap
Since the purpose of fractionation is to break up the
solid unity of a fixed pattern rather than to provide a
descriptive analysis it does not matter if the fractions do
not cover the whole situation. It is enough that one has
something to work with. It is enough that one has a new
arrangement of information to provoke restructuring of
the original pattern.

For the same reason it does not matter if some of the
fractions overlap. It is much better to produce some
sort of fractionation no matter how impure than to sit
wondering how a pure fractionation can be made.

If the problem being considered was 'transport by bus'
the following fractionation might be made:
Choice of route
Frequency
Convenience
Number of people using the service
Number of people using the service at different times
Size of bus
Economics of use and cost
Alternative transport
Number of people who would have to use the bus and
number who would like to use it if it were running.

Clearly these fractions are not all separate but overlap to
a considerable extent, for instance convenience is a
matter of route, frequency and perhaps size of the bus.
Economics of use and cost include the number of
people using the service, size of bus and several other of
the fractions.

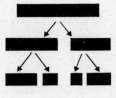

Two unit division

Whenever there is difficulty in dividing something into fractions it can be useful to adopt the *artificial* technique of division into two units or fractions. The two fractions so produced are themselves further divided into two more fractions and so on until one has a satisfactory number of fractions.

This technique is highly artificial and it can mean that several important features are quite overlooked. The advantage is that it is much easier to find two fractions than to find several. It is not a question of dividing something into two equal fractions for any two fractions will do no matter how unequal. Nor do the lines of division have to reveal natural fractions. The fractions may be very artificial and yet be useful.

Applied to the apple picking problem the two unit division might go as follows:

		delicate	damaging
	apple		damaged
		separate	finding
			density
apple picking			hold
problem		remove	jerk
	picking		to ground
		transport	container

The technique of two unit division is not so much a technique but a method for encouraging the fractionation of a situation.

Practice

1 Fractionation

The students are given a subject and asked to fractionate it. The subject may be a design project, a problem or any specific theme. Suggestions for subjects might include:

Unloading ships in harbour.
Restaurant meals.
Catching and marketing of fish.
Organization of a football league.
Building a bridge.
Newspapers.

The separate fractionation lists are collected from the
students. If there is time the results are analysed in
terms of the most popular fractions. If there is not time
then individual lists are read out and particularly
ingenious fractionations are commented upon. The
main purpose is to show the variety or the uniformity of
the approach.

2   Reassembly
From the fractionation lists obtained above (or from a
special session) are extracted small groups of two or
three fractions. These are then given to the students
who are asked to put them together again in an attempt
to generate a new way of looking at the situation.

3   Picking out fractions
Here the subject is presented to the students as a group.
They are asked to pick out fractions one after the other.
One student volunteers one fraction and then another
student follows with a further fraction. This continues
as long as suggestions are still coming in. It does not
matter if there is a considerable degree of overlap
between the suggested fractions. If there seems to be a
direct duplication this is pointed out to the person
making the suggestion and he is asked to say why he
thinks there is a difference. It does not matter whether
the difference is a very valid one or not so long as he
himself seems to think there is a difference.

4   Working backwards
This is as much a game as anything else. A list of

fractions is taken from a previous session with another group and the students are asked to try and guess what the subject was. Obvious references to the subject are deleted and substituted by the word 'blank'.

Another way in which this can be done is to give the students a list of five subjects only one of which is to be fractionated by each student. At the end some of the fraction lists are read out and the students have to decide which of the five original subjects a list refers to.

5   Two unit division
Here a subject is given to the students who are asked to carry out a two unit division on it. The end results are then compared. A quick comparison can be made between the first two units chosen by the different students. This can serve to show the variety of approaches used by the different students.

6   Sequential two unit division
A subject is given and then one student is asked to divide it into two units. Then another student is required to divide one of the units into two further units and so on. Unlike other practice sessions this one is not a matter of volunteering a solution but of being asked to provide one. The intention is to show that it is always possible to divide something into two units by picking out one unit and having the remainder as the other unit.

Summary
Fractionation may seem to be no more than straightforward analysis. The emphasis is however quite different. The aim is not to provide a complete or true breakdown of the situation into its component parts (as in analysis) but to provide material which can be used to stimulate restructuring of the original situation. The aim is restructuring not explanation. The

fractions do not have to be complete or natural for the emphasis is not on whether they are valid but on what they can bring about. The purpose of fractionation is to escape from the inhibiting unity of a fixed pattern to the more generative situation of several fractions.

Fractionation is a useful method for generating
alternative ways of looking at a situation. But it has
certain limitations. The fractions chosen are themselves
fixed patterns and usually standard patterns. The choice
of fractions is usually a vertical choice which follows the
most natural lines of division. The result is that the
fractions come together again to give a standard view of
the situation. Although fractionation makes it easier to
look at a situation in a different way the actual choice of
fractions limits the variety of alternatives that can be
generated. A simple square shape is shown opposite. If
one had to break this down into fractions one might
choose the fractions shown in any of the other figures.
Yet the choice of fraction will determine the shape that
can be made by reassembling the fractions differently.

The reversal method is more lateral in nature than the
fractionation one. It tends to produce more unusual
restructuring.

If you give someone an open-ended creative problem
there is great difficulty in getting started. There is
difficulty in moving at all. The person presented with
the problem seems to say, 'Where do I go, what do I
do?'. This was very obvious when I asked a group of
people to redesign some feature of the human body.
One obvious approach was to take some actual feature
as a starting point and then to modify it in some simple
way. Thus there were suggestions to increase the
number of arms or to lengthen the arms or to make
them more flexible.

Unless one is going to sit around waiting for inspiration
the most practical way to get moving is to work on what
one has. In a swimming race when the swimmers come
to turn at the end of the pool they kick hard against the
end to increase their speed. In the reversal method one
kicks hard against what is there and fixed in order to
move away in the opposite direction.

Wherever a direction is indicated then the opposite direction is equally well defined. If you go towards New York you are going away from London (or whatever other place you started from). Whenever there is action then the opposite action is indicated. If you are filling a bath full of water then the opposite action is to empty the bath. If something is happening over time then one merely runs the time scale backwards in order to find the reverse process. This is rather like running a cine film backwards. Whenever there is a one way relationship between two parties the situation can be reversed by changing the direction of this relationship. If a person is supposed to obey the government then the reversal would imply that the government ought to obey a person (or people).

In the reversal method one takes things as they are and then turns them round, inside out, upside down, back to front. Then one sees what happens. It is a provocative rearrangement of information. You make water run uphill instead of downhill. Instead of driving a car the car leads you.

Different types of reversal
There are usually several different ways in which one can 'reverse' a given situation. There is no one correct way. Nor should there be any search for some true reversal. Any sort of reversal will do.

For instance if the situation is: 'a policeman organizing traffic' then the following reversals might be made:
The traffic organizes (controls) the policeman.
The policeman disorganizes the traffic.

Which of these reversals is the better one? Either will do. It is impossible to say which arrangement will be the more useful until it has proved so. It is not a matter of choosing the more reasonable reversal or the more

143 of the reversal method

unreasonable one. One is searching for alternatives, for change, for provocative arrangements of information.

In lateral thinking one is not looking for the right answer but for a different arrangement of information which will provoke a different way of looking at the situation.

The purpose of the reversal procedure
Very often the reversal procedure leads to a way of looking at the situation that is obviously wrong or ridiculous. What then is the point of doing it?

● One uses the reversal procedure in order to escape from the absolute necessity to look at the situation in the standard way. It does not matter whether the new way makes sense or not for once one escapes then it becomes easier to move in other directions as well.
● By disrupting the original way of looking at the situation one frees information that can come together in a new way.
● To overcome the terror of being wrong, of taking a step that is not fully justified.
● The main purpose is provocative. By making the reversal one moves to a new position. Then one sees what happens.
● Occasionally the reversed approach is useful in itself.

With the policeman situation the first reversal supposed that the traffic was controlling the policeman. This would lead to consideration of the demand for more policemen as traffic became more complex, the need for redistribution of policemen according to traffic conditions. It would make one realize that in fact the traffic does actually control the policeman since his behaviour depends on the traffic build up in different roads. How quickly does he react to this? How sensitive is he to this? How well informed can he be of this? Since the traffic is controlling the policeman who is controlling the traffic why not organize things so that the traffic controlled itself?

The second reversal in the policeman situation supposed that the policeman was disorganizing the traffic. This would lead to a consideration of whether natural flow, traffic lights or a policeman was most efficient. If a policeman was more efficient than the lights what was the added factor—could this be built into the lights? Was it perhaps easier for the traffic to adjust to fixed patterns of direction rather than the unpredictable reactions of the policeman?

•

A flock of sheep were moving slowly down a country lane which was bounded by high banks. A motorist in a hurry came up behind the flock and urged the shepherd to move his sheep to the side so that the car could drive through. The shepherd refused since he could not be sure of keeping all the sheep out of the way of the car in such a narrow lane. Instead he reversed the situation. He told the car to stop and then he quietly turned the flock round and drove it back past the stationary car.

In Aesop's fable the water in the jug was at too low a level for the bird to drink. The bird was thinking of taking water out of the jug but instead he thought of putting something in. So he dropped pebbles into the jug until the level of water rose high enough for him to drink.

The duchess was much overweight. Physician after physician tried to reduce her weight by putting her on a near starvation diet and each physician was in turn dismissed on account of the unpleasantness of the diet. At last there came a physician who fussed over the good lady. Unlike the others he told her that she was not eating enough to sustain her huge body. He recommended that she drink a glass of sweetened milk half-an-hour before all meals (which of course reduced her appetite very much).

The rich man wanted his daughter to marry the richest of her suitors. But the daughter was in love with a poor student. So she went to her father and said that she wanted to marry the richest of her suitors but how could they tell which was the richest. It would be no use asking them to show their wealth by giving a present since it would be easy to borrow money for this purpose if the daughter was to be the prize. Instead she suggested that her father should give a present of money to each of the suitors. Then one would be able to tell how rich each was by the difference that the present of money made to their usual way of life. The father praised her for her wisdom and gave a present to each suitor. Whereupon the daughter eloped with her now enriched true love.

In each of these examples a simple reversal proved useful in itself. More often reversals are not especially useful in themselves but only in what they lead to. One ought to get into the habit of reversing situations and then seeing what happens. If nothing happens then there is no loss and there must be some gain in the challenge to the established way of looking at things.

Practice
1   Reversal and different types of reversal
A number of situations are presented to the students each of whom has to try reversing each situation in as many ways as possible. The results are collected and then the different types of reversal are listed. Comments are made on the more obvious types and also on the more ingenious types.
The same thing can also be done by giving out a subject and then asking for volunteer reversals of it, listing these down on a blackboard as they come in (and

supplementing them with one's own suggestions).

Possible subjects might be:
Teacher instructing students.
Street cleaner.
Milkman delivering milk.
Going on holiday.
Workers striking.
Shop assistants helping customers.
Comment
In some cases the reversal may seem utterly ridiculous.
This does not matter. It is just as useful to practise
being ridiculous as to practise reversal. In the above
examples (and the teacher can generate different ones) it
is not just a matter of reversing the given statement but
of reversing some aspect of the subject itself. For
instance 'going on holiday' can be reversed as 'holiday
coming to one'. On the other hand one might consider a
holiday as 'a change of scenery' and reverse this to a
holiday as 'complete uniformity of surroundings'.

2   What reversal leads to
Here one takes the situation and its reversal and sees
what the reversal leads to. This is best done in a general
classroom situation. The situation and its reversal are
offered to the class and volunteer suggestions are invited
as to lines of thought that the reversal might open up.
For instance the idea that 'holidays might involve
complete uniformity of surroundings' might lead to the
idea of freedom from decision, from stress, from having
to adapt.

To begin with it is not always easy to develop further
ideas from the reversed situation. That is why it is
better to do it in an open class situation rather than
require each student to work out something for himself.
Once the idea is grasped and everyone seems eager to
offer suggestions then each individual student can be

asked to reverse a situation and develop lines of thought
that arise from that reversal. In considering and
commenting on these at the end it is necessary to be able
to trace the line of development of an idea rather than
just have the end product. For that reason the students
ought to be encouraged to put down their train of
thought.

Brainstorming

What has been discussed so far includes the general
principles of lateral thinking and special techniques for
practising these principles and applying them.
Brainstorming is a formal *setting* for the use of lateral
thinking. In itself it is not a special technique but a
special setting which encourages the application of the
principles and techniques of lateral thinking while
providing a holiday from the rigidity of vertical
thinking.

The previous sections have described techniques that
could be used on one's own. The practice sessions have
involved a teacher-student interaction. Brainstorming is
a group activity. Nor does it require any teacher
intervention.

The main features of a brainstorming session are:
● Cross stimulation.
● Suspended judgment.
● The formality of the setting.

Cross stimulation
The fractionation technique and the reversal technique
are methods for getting ideas moving. One needs to
move to a new arrangement of information and then one
can carry on from there. The new arrangement of
information is a provocation which produces some
effect. In a brainstorming session the provocation is
supplied by the ideas of others. Since such ideas come
from outside one's own mind they can serve to stimulate
one's own ideas. Even if one misunderstands the idea it
can still be a useful stimulus. It often happens that an
idea may seem very obvious and trivial to one person
and yet it can combine with other ideas in someone
else's mind to produce something very original. In a
brainstorming session one gives out stimulation to
others and one receives it from others. Because the
different people taking part each tend to follow their

own lines of thought there is less danger of getting stuck with a particular way of looking at the situation.

During the brainstorming session the ideas are recorded by a notetaker and perhaps by a tape recorder as well. These ideas can then be played back at a later date in order to provide fresh stimulation. Although the ideas themselves are not new the context has changed so the old ideas can have a new stimulating effect.

Although the ideas in a brainstorming session are related to the problem under discussion they can still act as random stimuli for they can be far removed from the idea pattern of the person listening to them. The value of random stimulation is discussed in a later section.

Suspended judgment
The value of suspended judgment has been discussed in a previous section. The brainstorming session provides a formal opportunity for people to make suggestions that they would not otherwise dare make for fear of being laughed at. In a brainstorming session anything goes. No idea is too ridiculous to be put forward. It is important that no attempt at evaluation of ideas is made during the brainstorming session.
Attempts at evaluation might include such remarks as:
'That would never work because . . . . '
'But what would you do about . . . . '
'It is well-known that . . . . '
'That has already been tried and found to be no good.'
'How would you get that to . . . . '
'You are leaving a vital point out of consideration'.
'That is a silly, impractical idea.'
'That would be much too expensive.'
'No one would accept that.'

These are very natural remarks but if they are allowed then the brainstorming session is useless. Not only is one forbidden to evaluate the ideas of others but also one's own ideas. It is the job of the chairman of the session to stop any attempts at evaluation. He must make this quite clear at the start of the session. Thereafter he only need say; 'That is evaluation,' in order to put a stop to it.

The other type of evaluation which must be guarded against is the evaluation of the novelty of an idea. The object of a brainstorming session is to produce *effective* ideas. Usually this means new ideas otherwise one would not be holding the session. But the purpose of the session is not actually to find *new ideas*. During the session a long forgotten idea may be resurrected and found to be very effective.

The evaluation of novelty might include such remarks as:

'That is not new.'
'I remember reading about that some time ago.'
'That has already been tried in America.'
'That was the way it was done years ago.'
'I thought of that myself but threw it out.'
'What is so original about that idea?'

To counter such tendencies the chairman has to say, 'Never mind how new it is, let's have the idea and worry about its novelty later.'

Formality of the setting

Lateral thinking is an attitude of mind, a type of thinking. It is not a special technique much less a formal setting. Yet the value of a brainstorming session lies in the formality of the setting. The more formal the setting the better. The more formal the setting the more chance there is of informality in ideas within it. Most people are so steeped in vertical thinking habits that they feel very inhibited about lateral thinking. They do not like being wrong or ridiculous even though they might accept the generative value of this. The more *special* the brainstorming session is the more chance there is of the participants leaving their inhibitions outside. It is much easier to accept that 'anything goes' as a way of thinking in a brainstorming session than as a way of thinking in general.

Within this formal setting one can use all the other techniques that have been described so far for restructuring patterns and also those techniques which are yet to be described. One can try dividing things up into fractions and putting these together in new ways. One can try reversal. One does not have to apologize for it or even explain it to the others. The formality of the session gives one the licence to do what one likes with one's own thoughts without reference to the criticism of others.

Format for brainstorming session

● Size
There is no ideal size. Twelve people is a convenient number but a brainstorming session can work very well with as many as fifteen or as few as six. Less than six usually becomes an argument and with more than fifteen each person does not get enough opportunity to contribute. If there is a larger group then it can be broken down into smaller groups and notes can be compared at the end.

● Chairman
It is the chairman's job to guide the session without in any way controlling or directing it. He has the following duties:
1   The chairman stops people trying to evaluate or criticize the ideas of others.
2   The chairman sees that people do not all speak at once. (The chairman must also pick out someone who has been trying to say something but is always

outspoken by a more pushy character.) The chairman *does not* have to ask individuals to speak. They speak when they want to. Nor does he go round the circle asking each in turn for ideas. If however there is a prolonged silence the chairman may ask an individual for his thoughts on the matter.

3   The chairman sees that the notetaker has got an idea down. The chairman may find it necessary to repeat an idea or even to summarize an idea offered by a participant (this summary must be approved by the person whose idea it was.) The chairman may be asked to decide whether an idea is already on the list and so does not need listing again. If there is any doubt or the originator of the idea claims it to be different then it must be listed.

4   The chairman fills in gaps by offering suggestions himself. He may also call on the notetaker to read through the list of ideas already recorded.

5   The chairman can suggest different ways of tackling the problem and the use of different lateral thinking techniques for trying to generate different ways of looking at the problem (e.g. the chairman may say, 'Let's try turning this thing upside down.') Anyone else may of course make the same suggestions.

6   The chairman defines the central problem and keeps pulling people back to it. This is a difficult task since apparently irrelevant flights of fancy may be very generative and one certainly does not want to restrict people to the obvious view of the problem. As a guiding rule it may be said that any single flight of fancy is allowed but sustained divergence so that one comes to be considering a totally different problem is not allowed.

7   The chairman ends the session either at the end of a set time or if the session seems to be flagging – whichever is earlier. The chairman must not run the risk of boring people by extending the session indefinitely if it seems to be going well.

8   The chairman organizes the evaluation session and the listing of ideas.

● Notetaker
The function of the notetaker is to convert into a permanent list the many butterfly ideas that are put forward during the session. The task is a difficult one since the nebulous ideas offered must be reduced to manageable note form. Moreover the notes must not only make sense immediately after the session but some time later when the context is no longer so clear. The notetaker has to write fast for sometimes the ideas follow one another very rapidly. The notetaker can ask the chairman to hold things until he can catch up. The notetaker may also ask whether a particular summary of the idea is acceptable (e.g. shall we put this down as, 'More flexible traffic light system'?)

The notetaker must also assess whether an idea is new enough to be added to the list or whether it is already covered by a similar idea. If in doubt he should ask the chairman. It is better to put down duplicate ideas than leave out different ones for the duplicate ones can be removed later but the omitted ones are lost forever.

The notes must be in a form that is immediately readable for the chairman may ask for the list to be read out at any stage. It is not a matter of carefully transcribing shorthand some time after the end of the session.

It is useful to tape record a session as the playback may set off new ideas by repeating early ideas in a new context. Nevertheless even when the session is so recorded it is still essential to have a notetaker. At some time a summary list has to be made even of a tape and there is also the need to read out the list during the session.

- Time
  Thirty minutes is quite long enough for a session.
  Twenty minutes would be enough in many cases and
  forty-five minutes is an outside limit. It is better to stop
  while people are still full of ideas than to carry on until
  every last idea has been forced out. The temptation to
  carry on if the session is going well must be resisted.

- Warm up
  If the members of the group are not familiar with the
  technique (and perhaps even if they are) a ten minute
  warm up session is useful. This would deal with some
  very simple problem (bathtap design, bus tickets,
  telephone bells). The idea of this warm up session is to
  show the type of ideas that may be offered and to show
  that evaluation is excluded.

- Follow up
  After the main session is over the participants will
  continue to have ideas on the subject. These can be
  collected by asking each participant to send in a list of
  further ideas. If copying facilities are available then the
  list of ideas generated during the session can be sent to
  each participant with instructions to add any further
  ideas of his own on the bottom. '

Evaluation
As indicated above there is no attempt at evaluation
during the brainstorming session itself. Any tendency to
evaluate would kill spontaneity and convert the session
into one of critical analysis. Evaluation is carried out
later by the same group or even by another group. It is
important that some sort of evaluation is carried out
even if the problem is not a real one. It is the evaluation
session that makes a worthwhile activity of what would
otherwise be a frivolous exercise. In the evaluation
session the list of ideas is sifted to extract the useful ore.
The main points in the evaluation are as follows:

1   To pick out ideas which are directly useful.
2   To extract from ideas that are wrong or ridiculous the functional kernel of the idea which may be generalized in a useful way (e.g. in a brainstorming session considering the problem of rail transport one idea put forward was that trains should have tracks on their roofs so that when two trains met one could pass above the other. The functional idea here is fuller utilization of the same track or better use of carriage roofs.) The idea of using a magnet to pull apples from the trees would be considered as finding a means to bring apples en masse to the ground instead of picking them individually or as pretreatment of the apples in order to make them easy to pick.
3   To list functional ideas, new aspects of the problem, ways of considering the problem, additional factors to be taken into consideration. None of these are actual solutions to the problem but merely approaches.
4   To pick out those ideas which can be tried out with relative ease even though they may seem wrong at first sight.
5   To pick out those ideas which suggest that more information could be collected in certain areas.
6   To pick out those ideas which have in fact already been tried out.

At the end of the evaluation session there should be three lists:
● Ideas of immediate usefulness.
● Areas for further exploration.
● New approaches to the problem.

The evaluation session is not just a mechanical sorting for some creative effort is required to extract usefulness from ideas before they are discarded or to spot an idea which looks as if it ought to be discarded but can in fact be developed into something significant.

Formulation of the problem

While any problem can be the subject of a brainstorming session the way the problem is formulated can make a huge difference to the success with which it is tackled.

Too wide a statement of the problem may bring about a variety of ideas but they are so separated that they cannot interact to bring about that chain reaction of stimulation that is the basis of brainstorming. The statement of a problem as, 'Better traffic control', would be too wide.

Too narrow a statement of the problem restricts ideas so much that the session may end up generating ideas not about the problem itself but about some particular way of handling it. The statement of a problem as, 'To improve traffic lights', would not lead to ideas about traffic control by means other than traffic lights. It might not even lead to ideas on better traffic control by traffic lights for attention might focus on ease of manufacture, ease of maintenance and reliability of traffic lights quite apart from their functional importance.

It is the chairman's duty to state the problem at the beginning of the session and to repeat this statement frequently in the course of the session. If it should prove to have been stated badly then he—or anyone else in the group—can suggest a better way of stating it. A suitable statement of the problem mentioned above would be: 'Methods of improving traffic flow given the present arrangement of roads.'

Examples

Transcript 1

The following is a transcription of part of a brainstorming session that was considering the redesign of a teaspoon.

. . . A rubber spoon.

. . . I feel that the secondary function of a spoon which is that of transferring sugar from the basin to the cup has largely disappeared and that a teaspoon in the shape of an egg whisk would be much more efficient.

. . . (Put down egg-whisk.)

. . . And make it electrically driven.

. . . Incorporate a musical box for the aesthetic function.

. . . Have something like a pipette tube which you dip in the sugar with your finger over the top and transfer sugar in that way. Then the sugar would be provided with a dispersing agent so that you would entirely lose the pleasure of stirring.

. . . Going back to the egg whisk I think one ought to have a sort of screw thing, rather like an electrical swizzle stick. The axle would be hollow . . .

. . . (Can I interrupt here? You are beginning to tell us how you would make it and that is not the function of this session.)

. . . No, I am just describing what it looks like.

. . . (Could you describe it more simply?)

. . . A rotating spoon?

. . . No it's got a screw. You know, a propeller type screw.

. . . You push it up and down?

. . . No it's electric, you just press the button on the top.

. . . It seems to me this is too complicated. Now you have ordinary sugar tongs and each individual would have his own sugar tongs and would pick up a couple of lumps of sugar. The tongs have two ends and you could create turbulence just as easily as with a spoon.

. . . Doesn't this restrict you to lump sugar?

. . . Yes, small lumps. But you can still get the quantity of sugar you want.

. . . (What shall we put down there?)

. . . Tongs.

. . . What about something like those ashtrays which spin as you press them. We could have something that you placed over a cup and as you pressed it it opened

out to release some sugar and at the same time spun to
stir the sugar in.
. . . If there is so much fun stirring in sugar then perhaps
we ought to have some sort of inert sugar which people
who don't like sugar could use in order to enjoy stirring
in.
. . . A once off spoon made of sugar.
. . . A device which contains sugar and which is moved
up and down in the cup. But if you don't want sugar
you keep a gate closed.
. . . I would like to take up the idea of electricity but not
using a battery or anything like that but using the static
electricity present in the body.
. . . This idea of a screw. One could do it on the autogiro
principle. As the screw went up and down the fluid
would make it revolve.
. . . Like a spinning top.
. . . A vibrating table that would agitate everything on it
—whether you had sugar or not.
. . . What about a sugar impregnated stick.

Transcript 2
The session was attempting to discover a better design
for the windscreen wiper/washer function. Something
to prevent impairment of vision by an accumulation of
mud and/or water.

. . . A conventional windscreen wiper with water or
some other washing agent coming in through the arm of
the wiper itself instead of being sprayed onto the screen
from another point.
. . . A rotating centrifugal disc . . .
. . . Like on a ship ?
. . . Yes.
. . . How about doing away with the screen and just
having a very fast flow of air through which no particles
of dust or water could penetrate.
. . . A wiper that would move straight across the screen

from side to side or from top to bottom, the rate to be controlled by the driver.

. . . Have a liquid which makes the dirt transparent so you don't have to take it off.

. . . A screen that acts as a shutter and wipes itself clean as it revolves.

. . . An electrically heated screen that boils off the water.

. . . Radar control of the car itself.

. . . A high speed screen that ejected some liquid as it went up and wiped it off as it came down.

. . . Ultrasonics.

. . . Make mudflaps compulsory on all vehicles.

. . . Develop two types of magnet, one of which attracts water and the other attracts dirt and locate them on the bottom.

. . . Channel water off the roof of the cab and so make wipers less necessary.

. . . Have a liquid windscreen.

. . . How about a surface which is perpetually in motion.

. . . Vibration.

. . . Have a circular car with a windscreen that passes round it and through a washer on the way round.

. . . Windscreen wiper with jets in the wiper.

. . . (I think we have that down already as jets in the wiper arm itself.)

. . . Experiment with rotating sponges and brushes and things other than the conventional sweeper.

. . . Sheet of water flowing down the windscreen and get rid of the wipers altogether.

. . . (So far we have been trying to get rid of the wiper. Suppose we did not want to get rid of the wiper but just to improve it. Is there any way we could do things hydraulically?)

. . . A very high pressure jet of water that would dislodge the dirt and also provide volume for washing it away.

. . . Experiment with partial windscreen so that you don't actually look through glass, you look through a gap.

. . . 3, 6 or 8 or any number of wipers operating along the bottom or along the top and sides of the screen.

. . . Have two fairly conventional windscreens that go up and down alternately and pass through wipers as they go up and down.

. . . Have a rotating screen part of which went underneath where it was cleaned so you always had a fresh piece.

. . . Have a choice of washer tanks so you could vary the liquid according to the conditions – for instance using something special to wash off oil.

. . . A periscope so that you could see above the dirt.

. . . Have a venetian blind principle.

. . . Have a double thickness of glass with water in between. The front sheet would have small holes through which the water was constantly trickling.

. . . Some screen that would intercept most of the dirt before it reached the windscreen proper.

. . . Change the driving position. Turn round and drive from the back.

. . . Drive in tunnels.

. . . Television arrangement so that driver does not have to actually look out.

. . . An ordinary wiper with a variable speed which is automatically adjusted according to the speed of the car or the amount of light getting through the windscreen or something like that.

. . . Have a multilayer windscreen in which you just peel off the outer dirty layer.

. . . Have a soluble surface windscreen so that the water is constantly dissolving it and so keeping it clean.

. . . Have windscreen made of ice which is constantly melting and so keeping itself clean.

. . . You could just put a layer of the soluble stuff on before you went out.

Comment
The remarks within brackets were made by the

chairman. No attempt is made to distinguish the
remarks of the other participants. The nature of the
suggestions varies from the outright ridiculous to the
solid and sensible. It may also be seen how one idea
springs from another. There is very little attempt at
evaluation. Almost every remark contributes a new idea.

Practice

The classroom is divided up into groups of a suitable
size for a brainstorming session. Each group elects its
own chairman. If there is any difficulty about this then
the teacher makes a suggestion. The notetaker is also
selected in each group. It may be useful to have an
auxiliary notetaker who can relieve the first one halfway
through the session.

The general principles of the brainstorming session are
explained with emphasis on the following points:
1   No criticism or evaluation.
2   Say anything you like no matter how wrong or
ridiculous.
3   Do not try and develop ideas at length or make
speeches, a few words is enough.
4   Give the notetaker a chance to get things down.
5   Listen to the chairman.

A warm up problem is then given to each group and
they have a ten minute warm up session. At the end of
this session they go straight into the main session for
thirty minutes.

The teacher may sit in on the groups in turn. It is better
not to be too intrusive. Few comments are made at the
time but mental notes are kept for discussion afterwards.
The only thing which justifies an intervention is any
tendency to evaluate or criticize.

At the end of the sessions the groups come together

again. In turn the notetakers from each group read out
the list of ideas. The teacher may then comment as
follows.

1   Comments on the actual session stressing perhaps
the tendency to evaluate or the tendency to be too timid.
2   Comments on the lists of ideas. These could point
out the similarity of some of the ideas, the originality of
others.

3   Comments on the tone of the ideas. Some of the
suggestions may have been quite sensible others quite
ridiculous. If the suggestions do tend to be too solemn
the teacher might point out that at least some of the
suggestions during the sessions should be outrageous
enough to cause a laugh.
4   The teacher then adds some ideas and suggestions of
his own concerning the problems that have been
discussed.

In going through the lists of suggestions the teacher
may pick out some of the more outrageous ideas and
proceed to show how they can be useful. This is done by
extracting the functional principle of the idea and
developing it further.

The general impression that should be encouraged is
that the brainstorming session is a generative situation
in which one should not be too selfconscious. In
practice there is a tendency for some students to show
off and try and be deliberately humorous if they know
that their suggestions are to be read out to the assembled
class. One has to deal with that situation as best one can
without denying people the right to be outrageous. One
way is to ask the person to explain the idea further.

Suggested problems for use in brainstorming sessions
might include:
The design of money.

The lack of sufficient playgrounds.
The need for examinations.
Mining under the sea.
Providing enough television programmes for everyone
to see what they want to see.
Making the desert fertile.
Heating a house.

In each case what is being asked for is a way of doing it,
a better way of doing it, a new way of doing it. These
are merely suggestions and the teacher ought to be able
to generate further problems.

● Evaluation
Evaluation sessions should not be held on the same day
as the brainstorming sessions. The evaluation sessions
are best done in front of the whole class and each idea is
considered in turn for its direct or indirect usefulness.

One can have different categories into which each idea is
placed. These might be:
Directly useful.
Interesting approach.
For further examination.
Discard.

An alternative to this general evaluation is to write the
brainstorm lists on the blackboard a few items at a time
and get each student to evaluate the items with votes.
At the end the different evaluations can be compared by
seeing how many 'votes' each item gets.

In this context the evaluation session is a necessary part
of the brainstorm session but not an important part.
Evaluations tend to be critical analysis and vertical
thinking. Emphasis should be directed much more to
the brainstorm session itself than to the subsequent
evaluation.

It is important in any attempt at evaluation not to give the impression that the outrageous ideas were only of use in the brainstorming session but not of much practical use anywhere else. Such an impression would limit suggestions to the practical and the solemnly sensible which though worthwhile in themselves would never lead to new ideas. One of the most important functions of the evaluation session is to show that even the most outrageous suggestions can lead to useful ideas.

Summary
The brainstorming session is of value as a formal setting which encourages the use of lateral thinking. The brainstorming session has a value as a group activity in which there is a cross stimulation of ideas. Otherwise there is nothing special to a brainstorming session that could not be done outside it. Some people equate creative thinking with brainstorming. This is to equate a basic process with one relatively minor setting which encourages the use of that process. Perhaps the most important part of the brainstorming session is its formality. When one is first getting used to the idea of lateral thinking it is helpful to have some special setting in which to practise it. Later on there is less need for such a setting.

Analogies

In order to restructure a pattern, to look at a situation in a different way, to have new ideas, one must start having some ideas. The two problems of lateral thinking are:
- To get going, to get some movement, to start a train of thought.
- To escape the natural, obvious, cliché train of thought.

The various techniques described so far have all been concerned with generating some movement. So is the analogy technique.

In itself an analogy is a simple story or situation. It becomes an analogy only when it is compared to something else. The simple story or situation must be familiar. Its line of development must be familiar. There must be something happening or some process going on or some special type of relationship to observe. There must be some development either in the situation itself or at least in the way it is looked at. Boiling an egg is a simple operation, but there is development in it. The egg is placed in a special container and heated. In order to bring the heat into better contact with the egg a liquid is used. This liquid also serves to prevent the temperature from rising above a certain value. In the process the egg changes its nature. This change is a progressive one that is proportional to the amount of time the egg remains in this special situation. Different people have sharply different tastes about how far they want the process to go.

The important point about an analogy is that it has a 'life' of its own. This 'life' can be expressed directly in terms of the actual objects involved or it can be expressed in terms of the processes involved. One can talk of putting an egg into water in a saucepan and boiling it for four minutes until the white is hard but the yolk still quite runny. Or one can talk of the changing state of an object with time when that object is subjected to certain circumstances. Analogies are vehicles for relationships and processes. These relationships and processes are embodied in actual objects such as boiled eggs but the relationships and processes can be generalized to other situations.

The analogy does not have to be complicated or long. A simple activity may suffice. Butterfly collecting is a special hobby yet the processes involved can be generalized to many other situations (e.g. rarity, supply and demand; information and search procedures; beauty and rarity; interference with nature for one's own uses; classification).

Analogies are used to provide movement. The problem under consideration is related to the analogy and then the analogy is developed along its own lines of development. At each stage the development is transferred back to the original problem. Thus the problem is carried along with the analogy.
In mathematics one translates things into symbols and then deals with these symbols by means of various mathematical operations. One forgets all about the real meaning of the symbols. At the end the symbols are translated back and one finds out what has become of the original situation. The mathematical operation is a channel which directs the development of the original problem.

Analogies can be used in the same way. One can translate the problem into an analogy and then develop the analogy. At the end one translates back and sees what might have happened to the original problem. It is probably more useful to develop the two in parallel. What is happening in the analogy is transferred (as a process or relationship) to the actual problem.

For instance one might use the analogy of a snowball

rolling down a hill to investigate the spread of rumours. As the snowball rolls down the hill the further it goes the bigger it gets. (The more a rumour spreads the stronger it gets.) As the snowball gets bigger it picks up more and more new snow. (The more people who know the rumour the more people it gets passed onto.) But for the snowball to increase in size there must be snow. At this point one is not sure whether the size of the snowball is being compared to the number of people who know the rumour or the strength of the rumour. Does the snow on the ground correspond merely to people who can be influenced by the rumour or to people predisposed to believe this sort of rumour? One is already being forced by the analogy to look hard at the problem itself. A large snowball—perhaps an avalanche—can be very destructive but if one is forewarned one can get out of the way. (A rumour can also be destructive but can one get out of the way if forewarned, should one try to escape, to stop it, or to divert it?)

Using an analogy in this way is very different from arguing by analogy. In argument by analogy one supposes that because something happens in a certain way in the analogy then it must happen in the same way in the problem situation. The use of analogies in lateral thinking is completely different. As usual one is not trying to prove anything. Analogies are used as a method for generating further ideas.

Choosing an analogy
It might be thought that the method would only be of use if a particularly apt analogy was chosen. This is not so. The analogy does not have to fit all along. Sometimes it is better when it does not fit for then there is an effort to relate it to the problem and from this effort can arise new ways of looking at the problem. The analogy is a provocative device which is used to force a new way of

looking at the situation.

In general the analogies should deal with very concrete situations and very familiar ones. There should be a lot going on. And what is going on must be definite. The analogy does not have to be rich in processes or functions or relationships for these can be generated out of any sort of analogy by the way it is looked at.

The analogy does not even have to be a real life situation. It can be a story provided the development of that story is definite.

As an analogy for the problem of vertical thinking one might use the story of how monkeys are supposedly caught by burying a narrow mouthed jar of nuts in the ground. A monkey comes along, puts his paw into the jar and grabs a handful of nuts. But the mouth of the jar is of such a size that it will only admit an empty paw but not a clenched paw full of nuts. The monkey is unwilling to let go of the nuts and so he is trapped.

With vertical thinking one grasps the obvious way of looking at a situation because it has proved useful in the past. Once one has grasped it one is trapped because one is very reluctant to let go. What should the monkey do? Should he refuse to explore the jar? This would be a refusal to explore new situations. Should he deny that the nuts were attractive? It would be silly to deny the usefulness of something for fear of being harmed by it on some occasion. Would it be better if the monkey had not noticed the jar? To be protected by chance is a very poor form of protection. Presumably the best thing would be for the monkey to see the nuts, perhaps even grab them, then to realize that the nuts were trapping it, to let go of them, and to find another way of getting at the nuts—perhaps by digging up the jar and emptying it out. So the major danger in vertical thinking is not that

of being trapped by the obvious but of failing to realize
that one may be trapped by the obvious. It is not a
matter of avoiding vertical thinking but of using it and
at the same time being aware that it might be necessary
to escape from a particular way of looking at a situation.

Practice
1 Demonstration
In order to make clear what is wanted during the
practice sessions it is useful to start by taking a particular
problem, choosing an analogy, developing the analogy
and relating it to the problem all along. This could be
done on the blackboard. Suggestions from students
would be accepted but they would not be asked for.

2 Relating an analogy to the problem
The problem would be given to the class. The teacher
would develop an analogy on the blackboard and the
students would be asked to volunteer at each point a
suggestion as to how any particular development in the
analogy could be referred to the given problem.

3 Individual effort
Here the analogy would again be developed by the
teacher but this time the individual students would each
relate it to the problem, writing down their ideas on a
piece of paper. At the end these results would be
collected and comments of the following sort could be
made :
(1)   The variety of different ways in which the analogy
was related to the problem.
(2)   Consistency or lack of consistency in the
development of the problem (i.e. was a feature in the
analogy always referred to the same feature in the
problem or did it change. There is no special virtue in
consistency.)
(3)   Richness of development with every detail
translated from the analogy to the problem or poverty of

development when only the major points were transferred.

4   Functions, processes, relationships
Here an analogy is developed by the teacher in concrete terms. The students (working on their own) have to repeat the analogy but using general terms of process, function and relationship, in place of the concrete terms. This is an exercise in *abstracting* these things from analogies.
Possible analogies for this sort of abstraction might include:
Having a bath.
Frying potatoes.
Sending a letter.
Trying to untangle a ball of string.
Learning to swim.

5   Choosing analogies
A list of problems or situations would be given to the students who would be asked in open class to volunteer analogies which might be fitted to each of the listed problems. Any student who volunteered a suggestion would be asked to elaborate it briefly by showing how he would apply it to the problem.
Possible problems for this exercise might include:
Design a machine to give change.
Ways of making shopping easier.
Better clothes.
Ensuring adequate water supply for cities.
What to do with junked cars.

6   Set problem
A problem is given to the classroom and each student chooses his own analogy and works through it relating it to the problem. At the end the results are collected and commented upon. In the course of such comments one might compare the different types of analogy chosen.

One might also compare the different aspects of the problem that have been highlighted by the different analogies. There may be occasions when the same idea has been reached by completely different pathways.

7 Same problem, different analogies
The same problem is given to all the students but different students are assigned different analogies. This can be done as a group exercise. The students are divided into groups all of which are to consider the same problem. Each group however is given a different analogy. At the end of the session a spokesman for the group (equivalent to the notetaker in the brainstorming session) summarizes how the group related the analogy to the problem.
Suggested problem :
Finding the way in fog.
Suggested analogies :
A shortsighted person finding his way around.
A traveller in a strange country trying to find the railway station.
Looking for something that has been lost in the house (e.g. a ball of string).
Doing a crossword puzzle.

8 Same analogy, different problems
This can be carried out in the same way as the previous session, either on an individual basis or on a group basis. Different problems are set but in each case they must be related to the same analogy. At the end notes are compared to see how well the analogy has been fitted to the different problems.
Suggested analogy :
Trying to start a car on a cold winter morning.
Suggested problems :
How to tackle a difficult mathematical problem.
Rescuing a cat from a high ledge.
Fishing.
Getting tickets for a very popular football match.

Summary
Analogies offer a convenient method for getting going when one is trying to find new ways of looking at a situation instead of just waiting for inspiration. As with other lateral thinking techniques the important point is that one does not start moving only when one can see where one is going. One starts moving for the sake of moving and then sees what happens. An analogy is a convenient way of getting moving for analogies have a definite 'life' of their own. There is no attempt to use analogies to prove anything. They are only used as stimulation. The main usefulness of analogies is as vehicles for functions, processes, and relationships, which can then be transferred to the problem under consideration to help restructure it.

The most important feature of the mind as an information processing system is its ability to choose. This ability to choose arises directly from the mechanical behaviour of the mind as a self-maximizing memory system. Such a system has a limited area of attention. A limited area of attention can only settle on part of an information field. That part of the information field on which the limited attention area settles is thereby 'chosen' or 'selected'. The process is in fact a passive one but one can still talk of choice or selection. The behaviour of this limited attention area and the system mechanics underlying it are explained in detail elsewhere*.

'Attention area' refers to the part of a situation or problem that is attended to. 'Entry point' refers to the part of a problem or situation that is *first* attended to. An entry point is obviously the first area of attention and it may or may not be succeeded by others depending on the complexity of the situation.

From an insight restructuring point of view the choice of entry point is of the utmost importance. One could almost say that when no further information is added to the system that it is the choice of entry point which brings about insight restructuring. Why this is so follows directly from the mechanics of this type of information processing system*.

Patterns are established on the memory surface that is mind by the sequence of arrival of information. Once established these patterns have a 'natural' behaviour in so far as they tend to develop in certain ways, and to link up with other patterns. The purpose of lateral thinking is to restructure these patterns and arrange information to give new patterns.

1 field

2 pattern

3 development

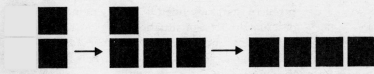

4 entry point

5 other pattern

The series of diagrams opposite illustrates the natural patterning behaviour of the memory surface of mind:

1 This shows the available information field.
2 Information is structured into a natural pattern.
3 The natural pattern has a natural line of development.
4 In developing the pattern there is a natural entry point from which one starts.
5 From the original information field only a limited area was selected by attention. Had the attention field been different then the pattern and its development would also have been different.

The choice of entry point is of huge importance because the historical sequence in which ideas follow one another can completely determine the final outcome even if the ideas themselves are the same. If you fill a bath using only the hot tap and then add the cold water at the end the bathroom will be thoroughly steamed up and the walls will be damp. If however you run some of the cold water in right at the beginning then there will be no steaming up and the walls will remain dry. Yet the actual amounts of hot and cold water will be exactly the same in each case.

The difference may be huge even if the actual ideas considered are the same but in practice a different entry point will usually mean a different train of ideas. A picture of a man with a stick in his hand followed by a picture of a dog running might suggest that the man is throwing sticks for the dog to retrieve. A picture of a dog running followed by a picture of the man with a stick in his hand might suggest that the man is chasing the dog out of his garden.

Entry point
Divide a triangle into three parts in such a way that the parts can be put together again to form a rectangle or a square.

The problem is quite a difficult one since the shape of the
triangle is not specified. You first have to choose a
triangle shape and then find out how it can be divided
up into three pieces that can be put together to give the
square or rectangle.

The solution to the problem is shown opposite. It is
obviously much easier to start with the square instead of
with the triangle which was suggested as the starting
point. There can be no doubt about the shape of a
square whereas the shape of a triangle (and to a lesser
extent of a rectangle) is variable. Since the three parts
have to fit together again to form a square one can solve
the problem by dividing up a square into three parts
that can be put together again to give a rectangle or a
triangle. Two ways of doing this are shown opposite.

In many children's books there is the sort of puzzle in
which are shown three fishermen whose lines have
gotten tangled up. At the bottom of the picture a fish is
shown attached to one of the lines. The problem is to
find which fisherman has caught the fish. The children
are supposed to follow the line down from the tip of the
fishing rod in order to find which line has the fish at the
end. This may involve one, two or three attempts since
the fish may be on any of the three lines. It is obviously
much easier to start at the other end and trace the line
upwards from the fish to the fisherman. That way there
need never be more than one attempt.

There is a simple problem which requires one to draw
the outline of a piece of cardboard which is so shaped
that with a single straight cut the piece can be divided
into four smaller pieces which are exactly alike in size,
shape and area. No folding is allowed.

The usual response to this problem is shown overleaf
with the percentage of people giving each type of

35% A impossible

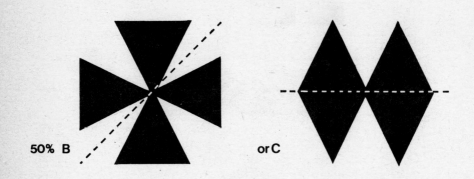

50% B   or C

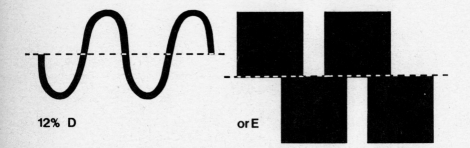

12% D   or E

3% F

answer. The solution given by groups B and C is obviously incorrect for a 'cut' has not thickness and so will divide the shape into two pieces and not four as required.

Answer D is correct. It is interesting that answer F is so rare for in hindsight it seems the easiest of them all (the explanation is that it is very difficult to think forward asymmetrically and in answer F the pieces are not all used in the same way). The important point of this problem, however, is that if one starts at the wrong end the problem is much easier to solve. Instead of trying to devise a shape that can be divided into four equal pieces one starts off with four equal pieces and clusters them around an imaginary cut. At first one might arrange them as shown overleaf but there is no difficulty in moving on to the next stage in which one shifts them along to give the solution.

To start at the wrong end and work backwards is quite a well-known problem solving technique. The reason why it is effective is that the line of thought may be quite different from what it would have been had one started at the beginning. There is no need to actually start at the solution end. It is convenient to do so since the solution is often clearly defined. But one can start at any point. If there is no obvious point then one must be generated.

Attention area
The entry point is the first attention area. Usually attention starts at this point but eventually covers the whole problem. Sometimes however important parts of the problem are completely left out. It is only when these parts are brought under attention that the problem can be solved.

In one of Sherlock Holmes' cases there was a large dog.

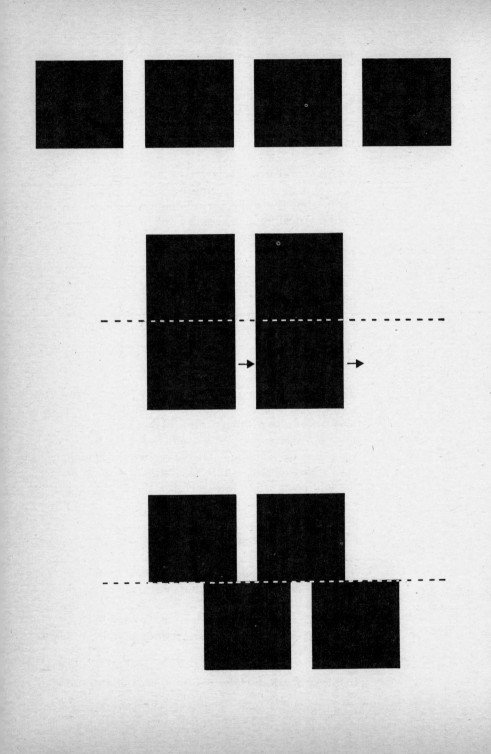

Dr Watson dismissed the dog as being of no importance because it had done nothing on the night of the crime. Sherlock Holmes pointed out that the great significance of the dog was precisely that it had done nothing. He shifted attention from the significance of what the dog might have done to the significance of the fact that it had done nothing. This meant that the criminal must have been known to the dog.

In Shakespeare's *Merchant of Venice* there comes the moment when Shylock demands the pound of flesh that is owed to him by the merchant as the result of a bargain. Shylock is outwitted by Portia who shifts attention from the flesh which is due to Shylock to the blood that must go with it. Since this is not part of the bargain Shylock could be charged with the serious offence of spilling blood. Thus by a shift of attention which brought into the problem something that would otherwise have been left out the problem was solved.

Two sets of circles are shown overleaf. In each case count up the number of solid circles as quickly as possible.

The obvious way to tackle this problem is to count the solid circles in each case. But when you come to the second set of circles it is much easier to shift attention to the open circles, find out the total number of these by multiplying the number of circles along one edge of the rectangle by the number along the other edge, and then subtract the small number of open circles from this total. The answer is the number of filled circles.

In a tennis tournament there are one hundred and eleven entrants. It is a singles knockout tournament and you as secretary have to arrange the matches. What is the minimum number of matches that would have to be arranged with this number of entrants?

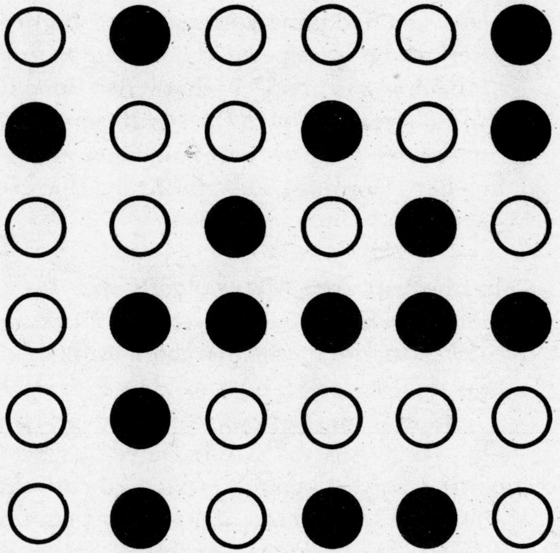

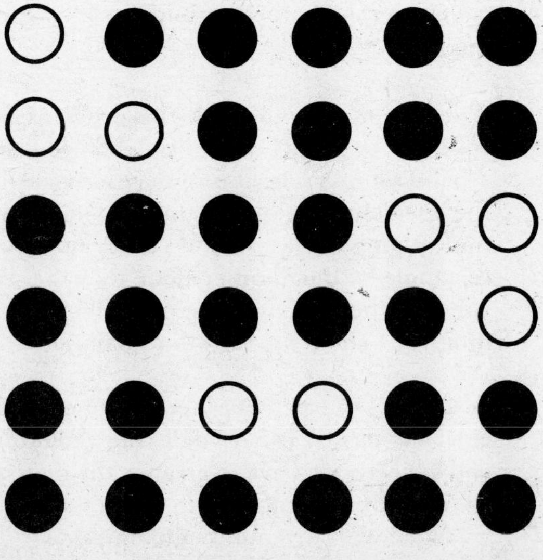

When faced with this problem most people draw little diagrams showing the actual pairings in each match and the number of byes. Others try and work it out by reference to 2^n (i.e. 4, 8, 16, 32 etc). In fact the answer is one hundred and ten matches and one can work this out at once without any complicated mathematics. To work it out one must shift attention from the winners of each match to the losers (in whom no one is usually very interested). Since there can only be one winner there must be one hundred and ten losers. Each loser can only lose once so there must be one hundred and ten matches.

In a sense this last problem could be regarded as an example of the usefulness of shifting the entry point except that the losers are usually never considered at all. Very often in a situation it is not just a matter of the order in which the parts are attended to but the choice of parts that are going to be attended to at all. If something is left out of consideration then it is very unlikely that it will ever come back in later on. Nor is there usually anything in what is being attended to that will indicate what has been left out.

For these reasons the choice of attention area can make a huge difference to the way the situation is looked at. To restructure the situation one may need no more than a slight shift in attention. On the other hand if there is no shift in attention it may be very difficult to look at the situation in a different way.

Rotation of attention
Since attention is basically a passive phenomenon it is no use just hoping that attention will flow in the right direction. One has to do something about it. Even though the process is passive one can still direct attention by providing a framework which will affect it. For instance you could decide that whenever you found

yourself staring at something then you would shift your gaze to a spot about two feet to the left of whatever you were staring at. After a while attention would automatically shift to that spot even though there was nothing there which attracted it. Attention follows the patterns set up in the mind not the external ones.

As with the reversal procedure one can deliberately turn away from what one would naturally pay attention to in order to see what happens if one paid attention to something else. For instance in the tennis tournament problem one might have said, 'I am trying to see how many matches there would have to be to produce one winner – instead of this let me see how many matches there would have to be to produce 110 losers.' This reversal procedure can work very well if there is a definite natural focus of attention in the situation.

Another method is to list the different features of the situation and then to proceed methodically through this list paying attention to each feature in turn. The important point here is not to feel that some features are so trivial that they do not merit any attention. The difficulty is that in any situation one can pick out as many features as one likes since the features reside not in the situation but in the way it is looked at.

Suppose one was considering the problem of homework. One might list the following features for attention in rotation:
Necessity for doing it (optional or required).
Time in which to do it.
Essential to course or reinforcing.
Travel time to get home.
Place to do it at home.
What else might be done instead.
Competing television programmes.
Routine or occasional.

Ability of father or mother to help.
Fast workers and slow workers.
Is one interested in what is done or the amount of time
spent doing it?
Frustration and annoyance of homework.
Homework as lessening the content or impact of
schoolwork.

Suppose the problem was one of getting rid of weeds.
The natural attention focus is the growth of weeds
which leads to methods for getting rid of them. But no
attention is paid to what happens after the weeds are
gone or to what would happen if the weeds were to stay.
Attention is on the weeds and getting rid of them. In a
recent experiment some strips in a field were sprayed
with the usual weedkiller and others left to grow weeds.
It was found that the yield of crops from the unsprayed
strips was in fact higher.

In a foot and mouth epidemic it is customary to burn
the corpses of infected animals if the soil is not deep
enough to bury them. But in the burning currents of hot
air rise and spread particles from the fire over a very
wide area. It is possible that such particles might be
infected with virus that has escaped the full heat of the
fire and so the disease might tend to spread. Here the
attention is on getting rid of the infected animals not on
the effect of the method used for getting rid of them.

A very useful drug in medicine was discovered when
someone noticed that when the drug was being used for
something quite different the patients always passed a
lot of urine. Since this was not the purpose of the
treatment no one paid any attention to it until someone
suddenly realized that here was a useful drug which
could make patients pass urine when this *was* the
purpose of treatment.

Practice

1   Identify entry points
An article discussing a particular problem is read out or
given to the students. They are asked to list possible
entry points for tackling the problem. They are also
asked to define the entry point used by the writer of the
article. Foı instance in an article on world hunger the
writer might have chosen the wastage of food in some
countries as his entry point, or he might have chosen
overpopulation or inefficient agriculture. From the
results the teacher lists the possible entry points that
have been suggested and adds other ones.

2   Entry points for assorted problems
A list of problems is written up on the blackboard and
the students are asked in open class to volunteer
different entry points for each of the problems. Each
student offering a suggestion is asked to elaborate it
briefly.

Possible problems might include:
The making of synthetic foods.
The acceptance of synthetic foods.
A better design for a sausage.
The problem of stray dogs.
An easy method for cleaning windows.

3   Same problem, different entry points
This could be done by individuals or as a group
activity. The same problem is set for all the groups but
each group is given a different entry point. At the end a
spokesman for each group discusses how they used the
entry point in each group. The point to watch here is
that the group really does use the entry point. There is
a temptation to consider the problem in the obvious way
and then just to connect up the entry point with this
obvious way.

Suggested problem :
A method for keeping rain off one while one is walking
in the street.
Suggested points of entry :
Bother of having to carry an umbrella.
Awkwardness of umbrellas when several people are
using them.
Why go out in the rain?
Why does getting wet matter?

4 Omitted information (story)

In telling a story one normally leaves out all the
information which is not essential for the development
of the story. But if one wants to examine the situation
itself rather than the way it has been described by
someone else then one has to try and put that
information back. One takes a story which may come
from a newspaper or may be a very well-known story. In
open class the students are asked for suggestions as to
what has been left out.
e.g. Jack and Jill went up the hill to fetch a pail of water.
Jack fell down and broke his crown and Jill came
tumbling after.
Was it on the way up or on the way back?
Was Jill hurt?
Why did Jill fall down anyway?
Why did Jack fall down?
Why were they going up hill to collect water?

5 Omitted information (picture)

Here a photograph or picture is used instead of a story.
One student examines the picture and describes it to the
classroom. Then each of the students draws a simple
version of what he thinks the described picture looks
like. From the nature of these drawings one can see the
information that was omitted in the description of the
picture. Another way of doing it is for a student to
describe the picture as before and for the rest of the

students to ask questions. Whenever a question can be answered from the picture then the student describing it could not have been paying attention to that part of the picture.

6   Further information
A picture is shown to the whole classroom. Each student writes down the information that he can get from that picture. At the end the results are collected and compared. The comparison between the person who extracts the *most* information and the person who extracts the *least* demonstrates how limited an area of attention may be.

7   Checklist
A problem is given and the students are asked to list all the different features through which they would like to rotate attention. This can be done in open class on a volunteer basis or by individual students with comparison of the lists at the end.
Suggested problems might include:
Alarm clocks that fail to wake one up.
Design for a bathtub.
Putting up a washing line.
Deciding where to build an airport.
Reducing noise from motorcycles and lorries.

8   Detective stories
With most detective stories there is difficulty in finding the criminal because certain factors are left out of consideration or the wrong entry point is chosen. The writer of a good detective story tries to bring about both these mistakes. The teacher devises a short detective story which contains enough clues to indicate who the criminal might be. The story is then read out to the class who each have to decide for themselves who the criminal is and why. The students should then be asked to write their own detective stories on these lines. These stories

in turn are read out to the class. For each story there is an assessment of how many students reach the right conclusion. The author of the story may be called upon to show how he has included enough clues to indicate the criminal.

Summary

Because of the nature of the self-maximizing memory system of the mind the entry point for considering a situation or a problem can make a big difference to the way it is structured. Usually the obvious entry point is chosen. Such an entry point is itself determined by the established pattern and so leads back to this. There is no way of telling which entry point is going to be best so one is usually content with the most obvious one. It is assumed that the choice of entry point does not matter since one will always arrive at the same conclusions. This is not so since the whole train of thought may be determined by the choice of entry point. It is useful to develop some skill in picking out and following different entry points.

The attention area is limited and includes much less information than is available. If something is left out of consideration then there is nothing which will make it come back into consideration at a later point. What is there does not usually indicate what is missing. Attention usually settles over the most obvious areas. A slight shift in attention may by itself restructure a situation. One tries deliberately to rotate attention over all parts of the problem especially those which do not seem to merit it.

The three ways of encouraging lateral thinking
discussed in this book are:
● Awareness of the principles of lateral thinking, the
need for lateral thinking, the rigidity of vertical thinking
patterns.
● The use of some definite technique which develops
the original pattern and may bring about restructuring.
● The deliberate alteration of circumstances so that
they can stimulate restructuring.

Most of the techniques discussed so far have worked
from within the idea. The idea has been developed
according to some routine process with the intention of
allowing the information to snap together again in a new
pattern. But instead of trying to work from within the
idea one can deliberately generate external stimulation
which then acts on the idea from outside. This is how
random stimulation works.

Some of the lateral methods discussed in this book have
not been very different from vertical methods though
the way they were used and the intention behind them
may have been different. The use of random stimulation
is fundamentally different from vertical thinking. With
vertical thinking one deals only with what is relevant.
In fact one spends most of one's time selecting out what
is relevant and what is not. With random stimulation
one uses any information whatsoever. No matter how
unrelated it may be no information is rejected as useless.
The more irrelevant the information the more useful it
may be.

Generating random inputs
The two main ways of bringing about random
stimulation are:
● Exposure.
● Formal generation.

Exposure

The division between exposure and formal generation
of random stimulation is only one of convenience. If one
actively puts oneself into a position where one is
subjected to random stimulation that is part exposure
and part formal generation. The following points may
serve to illustrate the way random stimulation can be
used.

1   Accepting and even welcoming random inputs.
Instead of shutting out something which does not
appear relevant one regards it as a random input and
pays it attention. This involves no further activity than
an attitude that notices what comes along.

2   Exposure to the ideas of others. In a brainstorming
session the ideas of others act as random inputs in the
sense that they do not have to follow one's own line of
thought even though they occupy the same field of
relevance. Listening to others even if one disagrees very
strongly with their ideas can provide useful input.

3   Exposure to ideas from completely different fields.
This sometimes goes under the heading of 'cross
disciplinary fertilization'. It means discussing a matter
with someone in a totally different field. For instance a
medical scientist might discuss systems behaviour with
a business analyst or with a fashion designer. One can
also listen to other people talking on their own subject.

4   Physical exposure to random stimulation. This may
involve wandering around an area which contains a
multitude of different objects, for instance a general
store like Woolworths or a toy shop. It may also mean
going along to an exhibition which has nothing to do
with the subject you are interested in.

The main point about the exposure method is to realize
that one is *never looking for anything*. One could go to an

exhibition to see if there was anything relevant. One
could discuss a problem with someone in another field
in order to hear their views on it. But that is *not* the
purpose. If one goes looking for something relevant
then one has preset ideas of relevance. And such preset
ideas of relevance can only arise from the current way of
looking at the situation. One wanders around with a
completely blank mind and waits for something to
catch one's attention. Even if nothing seems to catch
one's attention there is still no effort to find something
useful.

● Formal generation of random input
Because attention is a passive process even if one
wanders around an exhibition without looking for
anything relevant attention does tend to settle on items
which have some relevance to the established way of
looking at a situation. No matter how hard one tries to
resist doing so one is still exerting some selection. This
reduces the random nature of the input but still allows it
to be very effective. In order to use truly random inputs
one has to generate them deliberately. This seems
paradoxical in so far as a random input is supposed to
occur by chance. What one actually does is to set up a
formal process to produce chance events. Shaking a pair
of dice is such a situation. Three methods are suggested
below.

1   Use of a dictionary to provide a random word.
2   Formal selection of a book or journal in a library.
3   The use of some routine to select an object from the
surroundings (e.g. the nearest red object). The use of a
dictionary will be described in more detail further on in
this section. Formal selection of a book or journal
simply means that one makes a point of picking up a
journal from a particular position on the shelves no
matter what the journal may be. One opens it and reads
any one article in it no matter how remote these may
seem. One can do the same thing with a book. These are

but examples of how one can set up deliberate habits or routines in order to generate random inputs.

The effect of random stimulation

Why should random stimulation have any effect? Why should a totally unrelated piece of information help to bring about the restructuring of an established pattern?

Random stimulation only works because the mind functions as a self-maximizing memory system. In such a system there is a limited and *coherent* attention span.* This means that any two inputs cannot remain separate no matter how unconnected they are. Normally if there were two unconnected inputs one of them would be ignored and the other one would be attended to. But if both are deliberately held in attention (by deliberately arranging the setting) then a connection will eventually form between the two. At first there may be a rapid alternation of attention between the two items but soon the short term memory effect* will establish some sort of link.

In this type of system nothing can be truly irrelevant.

The established patterns on the memory surface are stable patterns. That does not mean that they do not change but that the pattern of change is stable. The flow of thought is stable. This equilibrium state is altered by the sudden inclusion of some new information.

Sometimes the new equilibrium state is very similar to the old one with a slight alteration to include the new information. At other times a complete restructuring comes about. There is a game in which plastic discs are placed within a frame one side of which is being forced inward by a spring. The pressure of this spring forces the plastic discs together to give a stable structure. Each player in turn removes a plastic disc. Usually the pattern shifts slightly to achieve a new equilibrium state. But sometimes there is a big change and the whole pattern is restructured. With a random input one is putting something in instead of taking it out but the shift in equilibrium occurs in the same way.

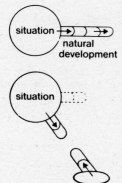

situation

natural development

situation

random stimulus

Random stimulation can work in two other ways. The random input can bring about a new entry point to the problem under consideration. The diagram opposite suggests a situation and the natural way this situation would develop. A random input is then added and a connection develops between the situation and the random input. As a result a new entry point is provided and the line of development of the original situation may be altered.

A random input can also work as an analogy. A single word from a dictionary provides a situation which has its own line of development. When this is related to the development of the problem being considered one has the analogy effect described in a previous chapter.

Random word stimulation

This is a practical and definite procedure in which the true random nature of the input is beyond doubt. If one is a purist one can use a table of random numbers to select a page in a dictionary. The number of a word on that page (counting down the page) can also be obtained from the table of random numbers. With less trouble one can simply think of two numbers and find the word that way. Or throw some dice. What one must not do is to open a dictionary and go through the pages until one finds a likely looking word. That would be selection and it would be useless from a random stimulation point of view.

The numbers 473–13 were given by a table of random numbers and using the Penguin English Dictionary the word located was: 'noose'. The problem under consideration was 'the housing shortage'. Over a timed three minute period the following ideas were generated.

noose–tightening noose–execution–what are the difficulties in executing a housing programme–what is the bottleneck, is it capital, labour or land?
noose tightens–things are going to get worse with the present rate of population increase.
noose–rope–suspension construction system–tent like houses but made of permanent materials–easily packed and erected–or on a large scale with several houses suspended from one framework–much lighter materials possible if walls did not have to support themselves and the roof.
noose–loop–adjustable loop–what about adjustable round houses which could be expanded as required–just uncoil the walls–no point in having houses too large to begin with because of heating problems, extra attention to walls and ceilings, furniture etc–but facility for slow stepwise expansion as need arises.
noose–snare–capture–capture a share of the labour

market – capture – people captured by home ownership due to difficulty in selling and complications – lack of mobility – houses as exchangeable units – classified into types – direct exchange of one type for similar type – or put one type into the pool and take out a similar type elsewhere.

Some of the above ideas may be useful, others may not. All of them could have been arrived at by straightforward vertical thinking but that does not mean that they would have been arrived at this way. As discussed before if an idea is tenable at all then it must be possible in hindsight to see how it could have been arrived at by logical means but this does not mean that it would have been arrived at in this way. Sometimes the link to the random word may be effected after the idea has come to mind rather than the random word stimulating the idea. Nevertheless the use of the random word has stimulated a large number of different ideas in a short period of time.

From this example may be seen the way the random word is used. Often the random word is used to generate further words which themselves link up with the problem being considered. Examples of this include: noose – execution – bottleneck; noose – rope – suspension; noose – snare – capture. A chain of ideas stretches out from the random word in order to effect a link with the problem. At times the functional properties of a noose were transferred to the problem: tightening noose, adjustable, round. The random word can be used in these and in many other ways. *There is no one correct way to use it.* In some cases a pun on the word may be used, or its opposite, or the word spelled slightly differently. The word is used in order to get things going – not to prove anything. Not even to prove that random word stimulation is useful.

● Time allowed
In the above example the time allowed was three
minutes. This is quite long enough to stimulate ideas. If
one sits around with a word long enough then it can
become boring. With practice and confidence three
minutes should be enough or at most five minutes.
What one must not do is to immediately look for
another random word at the end of the period because
this tends to set up a search routine in which one goes
through word after word until one finds a suitable one.
Suitable would only mean one that fits in with the
established view of the situation. If one wants to try
another word it should be on another occasion.
Knowing that one is going to move directly to another
word (and hopefully a better one) reduces the
effectiveness of the first word. Even after the end of the
fixed period further ideas will occur. One can note them
down. But there is no question of going through the
rest of the day desperately trying to extract the
maximum from the random word. One can get into the
habit of using a random word on a problem for three
minutes every day.

● Confidence
The most important factor in the successful use of
random stimulation is confidence. There is no sense of
urgency or effort but a quiet confidence that something
will emerge. It is difficult to build such confidence
because at first ideas will be slow to come. But as one
learns to handle random stimulation in the knowledge
that nothing can be irrelevant it becomes easier and
easier.

Practice
1   Relating a random word
A problem is stated and written out on the blackboard.
The students are then asked for suggestions of a number
up to the number of pages in a dictionary (e.g. a number

from 1 to 460) and then for another number to give the
position of the word on that page (e.g. 1 to 20). Using
a dictionary the corresponding word is located. The
word is written down together with its meaning (unless
the word is a very familiar one). The students are then
asked for suggestions as to how the word could be
related to the problem. To begin with the teacher may
have to make most of the suggestions himself until the
students get used to the process. Each suggestion is
elaborated briefly but no attempt is made to note down
the suggestions. The session goes on for 5–10 minutes.

Possible problems:
How to deal with the problem of shoplifting.
Increasing car safety.
A new design for windows to make them easier to open
and close without the danger of people falling out or
draughts.
New design for a lampshade.

Unless the teacher is fairly confident about his ability to
use *any* random words it might be better to use the list
given below rather than a dictionary. In this case the
class would be asked for a number from 1 to 20.

1	weed	11	tribe
2	rust	12	puppet
3	poor	13	nose
4	magnify	14	link
5	foam	15	drift
6	gold	16	duty
7	frame	17	portrait
8	hole	18	cheese
9	diagonal	19	chocolate
10	vacuum	20	coal

2   Same problem, different words
Here a problem is set but different random words are
used. Each student works on his own and makes notes of

how the word generates ideas about the problem. At the
end the results are collected. If there is time these are
analysed to see whether there is any consistency of
approach which depends on the random word used.
The same idea may have been reached in different ways
depending on the random word. If there is not much
time then some of the results are selected at random and
read out. One can also take the end idea in each chain of
thought and then ask the class to imagine what the
random word was in this particular case and the line of
thought that led to it. (e.g.) if the problem was 'holidays'
and the random word was 'turkey' a chain of thought
might run: turkey – special food – Christmas – special
holiday – more holidays with a special purpose. One
would just take the 'more holidays with a special
purpose' and ask what the random word might have been .

Two or three random words distributed among the
class would be enough. More would just be confusing.
The words can be taken from a dictionary or from the
list above.

Possible problems might include :
Clearing oil off a beach.
Weeding the garden.
Design of apparatus for rescuing people from a burning
building.
Making plastic sheet suitable for clothing (how would
one treat it to make it hang properly).

3   Same word, different problems
This may be done either as an individual practice
session or as an open class session. A random word is
selected and then each student is given one out of two or
three chosen problems. The student works to relate that
random word to the problem he has been assigned. At
the end the results are compared to show the different
uses of the same word.

As an open class session three problems are listed. The random word is then related to each of the three problems in turn. Five minutes are spent on each problem. Suggestions are volunteered by the students and the teacher adds his own whenever there is a pause. It is better if the three problems are not written up together for then some students might be thinking ahead to the next problem.

Possible random words:
drain
engine
cooking
leaf
Possible problems:
How to store information so that it is easily available.
How to spend less time learning a subject.
A device to help you climb trees.
Design for a better cinema.

4   Your own problems
The students each write down any problem they would like to tackle. They write it down in duplicate, put a name or number on each sheet, and give one copy to the teacher. This is to prevent a sudden change in the problem when the random word is given. A random word is then found (by page number etc suggested by the students to locate a word in the dictionary or just chosen by the teacher).

Before they hand in their results some of the students are asked by the teacher to describe to the rest of the class how they related the word to their own problem. In this type of session one can get an idea of how the same random word can be of use in many different situations. If some students find that they cannot make any progress at all then the teacher goes through the problem with them showing how the random word might be used in each case.

Possible random words:
scrambled eggs
screwdriver
bomb
doorhandle

5   Random objects
The objects are not random to the teacher who selects
them but to the students to whom they are presented.
The advantage of an object over a word is that an actual
object can be looked at in many more ways than the
word describing that object. One should be able to
imagine an object in just as much detail but in practice
one does not and the function of the object tends to
swamp the other features. A problem is given to the
students and then the random object is presented. This
can either be run as an open class session with the
students making suggestions as to how the object may
be related to the problem or it can be done on an
individual basis with comment on the results or
individual students describing their own results.

Possible objects include:
a shoe
a tube of toothpaste
a newspaper
an apple
a sponge
a glass of water
Possible problems might include:
Learning how to swim.
A new design for clocks.
A device for getting handicapped people in and out of
bed.
Unblocking a drain.

Summary
If one only works from within an established pattern
then one tends to follow its natural line of development

and is unlikely to restructure the pattern. Usually one waits patiently for chance circumstance to provide information that will trigger off an insight restructuring. With random stimulation one deliberately mixes in an unconnected piece of information in order to disturb the original pattern. From this disturbance may come a restructuring of the pattern or at least a new line of development. For the random input to be effective there must be no selection about it for as soon as there is selection there is relevance and the disturbing effect of the random input is reduced. Random stimulation is a provocation. Because of the way the mind works any stimulus whatsoever can be found to develop a connection with any other.

Division

A limited and coherent attention span arises directly
from the mechanics of the self-maximizing memory
surface that is mind. This limited attention span means
that one only reacts to a bit of the total environment.
Over a period of time one bit may be attended to after
another until the total environment is covered.

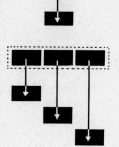

In effect the total, continuous and overwhelming
environment is divided up into separate attention areas.
The process may involve picking out a single attention
area or it may involve dividing up the environment into
a number of attention areas. This is shown in the
diagram opposite. There is no basic difference between
the two processes except that one covers the whole field
and the other does not.

Although this process arises directly from the mechanics
of the system it has several very useful advantages.

1   It means that some part of the environment can be
reacted to specifically. Thus if the total environment
contained something useful and something dangerous
one could react differently to each part.

2   It means that new and unfamiliar environments can
be dealt with by picking out features that are familiar.
Eventually the situation is explained in terms of such
familiar parts.

3   It means that the separate parts can be moved
around and combined in different ways to produce
effects that are not available in the environment.

4   It makes communication possible because a
situation can be described bit by bit instead of as a
whole.

Separation into units, selection of units, and

combination of units in different ways together provide
a very powerful information processing system. All
these functions follow directly from the mechanism of
mind.*

● Reassembly
The previous diagram shows how units can be created
by dividing up a total situation. Units can however also
be created by putting together other units to form a new
one that is then treated as a complete unit.

Words, names, labels
When a unit is obtained by dividing up a total situation
or by putting together other units it is convenient to
'fix' that unit by giving it a separate name. The name is
separate and unique to itself. The name establishes it as
a pattern in its own right instead of just being part of
another pattern. Having a name gives a unit much
greater mobility since it now becomes more sharply
divided off from its neighbours and comes to exist on
its own. A name is especially useful for combining
different units together to give a new one. The new unit
only exists in so far as it is given a name. Without that
name it would dissolve back into its separate parts.

The use of names for units is essential for
communication. Names make it possible to transfer a
complex situation a piece at a time.

To be of any use in communication the names must be
fixed and permanent. Once a name is assigned to a unit
then the shape of that unit is 'frozen' because the name
itself does not change. This fixity of name is vital for
communication and it is also useful for understanding a
situation. In understanding however one does not
actually have to use names though most people find it
convenient to do so.

Myths

Myths are patterns which first arise in the mind. Once these patterns have formed something may be found in the environment which justifies them or else they dictate the way the environment is looked at and so achieve a pseudojustification. Once one has names then one can do things to the names themselves and so produce more names. Thus if one has a word one can produce a word with an opposite meaning merely by adding 'un'. One can then look around to see what this new word fits or use it anyway whether it represents anything or not. Similarly once one has two words one can put them together to give a third word which is a combination of the other two. Both these processes are shown opposite. These new units are created on the word level rather than derived from the environment. Yet these myth words are treated in exactly the same way as ordinary words which do refer to actual things in the environment. Instead of the myth word following something in the environment the myth word comes first and actually 'produces' something in the environment (by dictating the way one looks at something). Both sorts of words have the same degree of permanence and reality. Both are treated in exactly the same way.

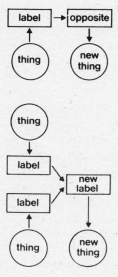

Limitations of the naming system

The great practical advantage of the named unit system is its permanence and the great practical disadvantage of the named unit system is its permanence.

Names, labels, words, are themselves fixed and permanent. Hence the units which have been taken over by these names have also to be fixed and permanent. Hence the patterns which are arrangements of such units tend also to be fixed and permanent.

The major disadvantage is that a named unit which

might have been very convenient at one time may no longer be convenient, indeed it may be restricting. The named assemblies of units (which are called concepts) are even more restricting because they impose a rigid way of looking at a situation. When there is a famine in a rice eating country and maize is sent in by other countries the starving people prefer to starve. Such is the rigidity of the concept 'maize is food for animals'.

Even without a name a concept would be fixed by repeated use and growing familiarity. Putting a label on it accelerates the process.

Some of the limitations that arise from this named unit process are outlined below.

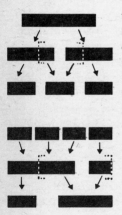

1   A division at a point of convenience produces two units which then become established and named. Subsequently it may be more convenient to divide the original situation into three units. This is shown in the top figure opposite. The establishment of the new units is very difficult as it means carving bits out of the previous units and putting them together so that they form a new unit rather than revert back to the old units.

2   The lower figure opposite shows how an assembly of units becomes established as a new unit. If it becomes more convenient to change this assembly so that it includes some new units but excludes some old ones this is very difficult.

3   When a unit is separated out and named it is difficult to realize that it is part of a whole.

4   When an assembly of units is given an overall name it may be difficult to realize that it is made up of parts.

5   When a division has been made it is difficult to bridge across that division. If a process has been cut at

some point and what goes before that point is called
'cause' and what comes after is called 'effect' then it is
difficult to bridge across the point and call the whole
thing 'change'.

This is not a comprehensive list by any means. What
is implied is that if units have been cut out and
assembled in various ways which are then fixed by
labels it becomes very difficult to use different units or
different ways of putting them together.

Polarization

It is easier to establish two completely different patterns
than to change an established pattern. If a new pattern
is only slightly different then it will shift towards the
established pattern. There is a tendency for established
patterns to 'mop up' similar patterns which are treated
as a repetition of the standard pattern. This results in a
distortion of the information that is actually presented.
The pattern that would have been established by the
information is shifted towards an established pattern. If
there are two established patterns then the shift may be
towards one or other. If the two established patterns are
opposite 'poles' in any sense then this shifting moves
the new pattern towards one or other pole.

It is like having two wooden boxes side by side into
which one is putting ping-pong balls. The balls have to
go into one box or the other. A ball will not balance on
the division between the two boxes. If the edges of the
boxes are sloping then the ball may be moved quite a
long way. The process is suggested by the diagram
opposite.

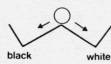

black   white

If one of the boxes is labelled 'black balls' and the other
one 'white balls' then each ball is dropped into the
appropriate box depending on whether it is black or
white. If there are any grey balls then some sort of

decision has to be made as to whether they go into the black box or the white box. Once the decision is made the balls go into the white box just as if they were white or into the black box just as if they were black. The apparent nature of the ball has been shifted to make it fit in with the established pattern.

A whole series of boxes might be imagined, each with its own label. As each item came along it would be put into whichever box had the most appropriate label. It would not matter if this most appropriate label was not really very appropriate. There is a shift to fit in with whatever labels are available. Once the shift has been made then it is impossible to tell that the item in the box is any different from the other items in the box.

In order to find an appropriate box for any item that does not fit readily into any available box one can do two things. One can concentrate on those points which show that it ought to fit into one box. Or one can concentrate on those points which show that it should *not* fit into a particular box. Thus with the grey ping-pong balls one might have said, 'Grey is almost white therefore it fits into the white box' or one might have said 'Black is a true absence of any colour therefore the grey ball cannot go into the black box'.

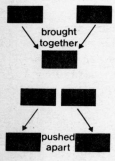

If two things are similar one could notice the points of similarity and say the two things are the same or one could notice the points of difference and say the two things are different. The two things would be shifted together to be similar or shifted apart to be different. Either way there would be some shift as suggested in the diagram.

Similarly when there is an established label a new item is either pushed right under that label or else pushed right out. In a community that is sharply divided into 'us' and 'them' any stranger who happens along is assessed as to whether he is 'one of us' or 'one of them'.

Probably the stranger has a mixture of characteristics which would make him fit either group. But whichever way the decision goes his characteristics are at once assumed to have changed so that they match exactly the characteristics of the label. The stranger is pushed towards one or other pole. He can not remain in between any more than the needle of a compass can remain undecided when a magnet is brought near to it.

From a practical point of view this polarizing system is very effective. What it means is that one can establish a few major categories and then push everything into one or other of these. Instead of having to assess everything in detail and then decide how one is going to react one merely assesses whether it fits into one category or another. This is not even a matter of exact fit but of pushing it one way or another. Once the thing has been pushed into a category then reaction is easy since the categories are established and so is the reaction to them.

In exploring a new situation one might have two categories: 'good to eat' and 'bad to eat'. This is sufficient. Anything examined can be pushed one way or the other. There is no need to have any subtle distinctions. Such distinctions as: 'tastes nasty but is good for one', or 'good to eat but makes one very thirsty', or 'tastes good but is poisonous', 'not known but worth trying' are excluded.

● New categories
At what point does a new category arise? At what point does one decide that the item will not fit into any of the boxes and so create a new box? At what point does one decide that grey ping-pong balls would go into a special box marked 'grey'? At what point is it decided that the

stranger is neither 'we' nor 'they' but something else?
The danger of polarization is that things can be shifted
around so much that there never comes a point when a
new category *has to be created*. Nor is there any
indication as to how many established categories there
should be.
One can get by with very few categories.

The dangers of the polarizing tendency may now be
summarized:
● Once established the categories become permanent.
● New information is altered so that it fits an
established category. Once it has done so there is no
indication that it is any different from anything else
under that category.
● At no point is it ever *essential* to create new
categories. One can get by with very few categories.
● The fewer the categories the greater the degree of
shift.

Lateral thinking
There is no question that the named unit system is
highly effective. There is no question that the polarizing
properties of this system make it possible to react with
very little information. The whole information
processing system that arises from the basic mechanism
of mind is immensely useful. The disadvantages
mentioned above are minor ones compared to the
usefulness of the system. But the disadvantages do
exist. Moreover they are inseparable from the nature of
the system. So one uses the system to its full
effectiveness but at the same time realizes the errors and
tries to do something about them.

The major limitation of the named unit system is the
rigidity of the labels. Once established the labels are
fixed. The labels alter the incoming information instead
of the incoming information altering the labels.

The aim of lateral thinking is to break out of cliché patterns and rigid labels are a perfect example of cliché patterns. In order to escape from these labels one can do three things:

● Challenge the labels.
● Try and do without them.
● Establish new labels.

● Challenging the labels
Why am I using this label?
What does it really mean?
Is it essential?
Am I just using it as a convenient cliché?
Why do I have to accept that label used by other people?

As it implies challenging a label means a direct challenge to the use of a label, a word, or a name. It does not mean that one disagrees with its use or that one has any better alternative. It just means that one is not prepared to accept the cliché label without challenging it.

It is not a matter of seeking justification for the label so that one can continue to use it. One continues to challenge the label all the time even when one is using it.

● Trying to do without labels
Whenever units are assembled together and given a new name or label this becomes so easily established that one tends to forget what lies underneath the label. By abolishing the label one can rediscover what there is underneath. One may find much of use that was hitherto hidden. One may find that there is very little of importance even though the label itself seemed to be important. One may find that the label is indeed useful but that it needs to be changed to bring it up to date.

By abolishing the label one abolishes the cliché

convenience of the label. If one is writing or speaking
one tries to proceed without the cliché convenience of
that label – without that label. Whenever one comes to a
point when one would normally use the label one has to
find a way of doing without it. This may involve finding
another way of looking at things and this of course is the
aim of lateral thinking. It is not much use substituting
some phrase instead of the label but it is still of some
use because the phrase can interact with other things in
the way a fixed label cannot.

A simple example of trying to do without a label would
be rewriting a very personal piece in which 'I' occurred
all the time. In rewriting it to avoid the use of 'I' one
would find that many things would have happened
anyway and that the personal involvement was much
less than had seemed.

It is not only in discussing a situation that one tries to
do without a particular label but also in looking at a
situation. Using the label 'mob' it is easy to develop a
certain line of thought but if one has to do without this
label then one might be able to look at the situation in a
different way. One tries to see things as they actually
are and not in terms of labels.

● Establishing new labels
It may seem paradoxical to establish a new label in
order to escape the harmful effects of labels. The
purpose of establishing a new label is however to escape
the distorting effects of the old labels. The polarizing
effect tends to shift information into established
categories. The fewer the categories the greater the
shift and distortion. By establishing a new category one
can accept information with less distortion. So one
establishes a new label in order to protect incoming
information from the polarizing effect of already
established labels.

Established labels tend to build around themselves meanings, contexts and lines of development. Even if one wants to use an idea that would fit under an existing label it might be better not to put it there if one wants to develop the idea in a new way. For instance lateral thinking does overlap with what some people understand by creative thinking. But because creative thinking is surrounded by a whole complex of meanings including artistic expression, talent, sensitivity, inspiration etc it is far better to establish lateral thinking as a separate idea if one wishes to regard it as a deliberate way of using information. Similarly the word 'patriotism' is so surrounded by heroics and duty and virtue and 'my country right or wrong' that one has to regard it as either very honourable or very dangerous. If one wants to encourage national spirit in terms of one country among others and in terms of individual culture and in terms of economic growth, then one needs a new label.

Practice

1   Challenging labels

This is rather similar to the 'why' technique described in a previous section. When one challenges a name, a label, or a concept one is *not asking for the term to be defined*. One is questioning the use of the term as a term not asking for its justification or explanation.

An article is taken from a newspaper or magazine and read out to the students. If there are enough copies they can be asked to read it for themselves. The task is to pick out certain labels which seem to be used too glibly. Each of these labels is underlined. It may be a label or a concept that is fundamental to the whole argument or it may be a label that is used very often. For instance in an article on management the labels picked out might include 'productivity', 'profitability', 'coordination'. Each student makes a list of such cliché words and at the end the lists are compared and discussed. The discussion is focused on how these labels are being used in too convenient a fashion. The point is not that the labels are right or wrong but that it is too convenient to write 'profitability'whenever one has to justify something. In another article the cliché words might be 'justice', 'equality', 'human rights'. In addition to discussing why the label is being used too glibly one also discusses the danger of using labels in this way.

2   Labels and discussion
Two students are asked to debate a subject while the rest of the students listen. At the end the other students comment on the use of labels during the discussion. It is enough that the students become aware of the easy use of labels. It is not a matter of deciding whether the label was justified nor a matter of commenting on debating techniques.

Possible subjects for such a discussion might include:
Are women as creative as men?
How far is obedience a good thing?
One should only learn subjects which are going to be immediately useful.
If you don't get what you want you should go on trying.
Parents should help children with their homework.
Children should dress as they like at school.
Some people are different from others.

3   Dropping labels
Here it is a matter of seeing how well one can do without a particular name or label or concept. The label is dropped completely and the article is rewritten without the use of that label. It is convenient to do this with newspaper articles that make much use of some particular label. In commenting on the result the teacher notes whether dropping the label has caused the thing to be looked at in a different way or whether the label has been replaced by a cliché phrase instead.

4   Dropping labels in discussion

Here one student is asked to discuss a subject. Then
another student is asked to explain what the first
student has said but without using some particular label
used by the first student. This type of thing can also be
done with a debate between students with both sides
forbidden to use some label. It can also be done with
only one of the sides forbidden to use the label.

Possible subjects for discussion
War (with label of fighting dropped).
Car racing (with label of fast, speedy, etc dropped).
Walking in the rain (with label of wet dropped).
School (with label of teaching dropped).
Police (with label of law dropped).

5   Rephrasing

Instead of dropping a concept label in the course of a
discussion or rewriting an article one practises doing it
with single sentences. This is rather more simple to do
than the previous exercise and it can be very useful
practice. The teacher selects a series of sentences which
may be taken from newspapers or just made up. The
sentences are read out or written up on the board. The
label which is to be dropped is underlined. The students
can then offer suggestions in open class as to how the
sentence could be rephrased without that word.
Alternatively they can each produce a version of the
sentence and at the end the different versions can be
compared. The important point with this exercise is
that the meaning must be kept as intact as possible.

The type of sentence which could be used is as follows:
Children should be as *tidy* as possible in their
homework.
Everyone has the right to *equal opportunity* in
education.
In a democracy government is by the *will* of the people.

If a thief is caught *stealing* he may be sent to prison.
Strawberry ice cream *tastes* better than vanilla.
If you drop a *plate* on the floor it will break.

The difficulty with this type of exercise is that very
often one simply gets synonoyms. Thus in the above
examples one might well get 'careful', or 'neat' instead
of the word 'tidy'. One cannot really refuse to accept
synonyms for the dividing line is very difficult between
what is a genuine synonym and what is a different way
of looking at the situation. So one accepts synonyms
but goes on further and asks for further ways of putting
things. Instead of refusing them one tries to exhaust
synonyms.

6 Headlines

This is very similar to the previous exercise. Instead of
sentences a series of headlines are taken from the
newspapers. The task is to rephrase the entire headline
so that no one word is the same as before and yet the
meaning is the same. It is necessary to choose headlines
which do not have specific labels in them. For instance
the headline 'Ribofillo wins Derby' would be difficult to
rephrase unless one were allowed to say, 'Favourite
triumphs in classic Epsom race' but this would imply
that one knew Ribofillo to be the favourite. One has to
allow some licence in this respect.

7 New labels

Since communication is so very important one does not
really want to encourage students to develop their own
special labels for things. One can however have an open
class session in which the students are asked to put
forward ideas which they feel are:
1 Improperly classified.
2 Left out by existing labels.

For instance someone may feel that a hovercraft is not

really an aeroplane or a car but something special.
Someone else might feel that 'guilty' and 'innocent' are
too sharp a division and that there should be room for a
person who is technically guilty but innocent as far as
intention goes (or technically innocent but actually
guilty).

Perhaps there ought to be a special label for something
which is not 'ugly' or 'beautiful' instead of having to
call it ordinary. Perhaps there ought to be a special label
to cover the phrase, 'the way you look at something'.
Perhaps there ought to be a special label for something
that was going well at the moment but was clearly
headed for disaster. Perhaps there ought to be a special
label for something that was not entirely an accident nor
was it entirely someone's fault but a mixture of the two.

Understanding the nature of lateral thinking and the
need for it is the first step towards using it. But
understanding and goodwill are not enough. The formal
routines suggested as methods of applying lateral
thinking are more practical but there is a great need for
something more definite, more simple, and more
universal. Some tool for applying lateral thinking just as
NO is a tool for applying logical thinking.

NO **and** PO

The concept of logical thinking is selection and this is
brought about by the processes of acceptance and
rejection. Rejection is the basis of logical thinking. The
rejection process is incorporated in the concept of the
negative. The negative is a judgment device. It is the
means whereby one rejects certain arrangements of
information. The negative is used to carry out judgment
and to indicate rejection. The concept of the negative is
crystallized into a definite language tool. This language
tool consists of the words no and not. Once one learns
the function and use of these words one has learned how
to use logical thinking. The whole concept of logical
thinking is concentrated in the use of this language tool.
Logic could be said to be the management of NO.

The concept of lateral thinking is insight restructuring
and this is brought about through the rearrangement of
information. Rearrangement is the basis of lateral
thinking and rearrangement means escape from the
rigid patterns established by experience. The
rearrangement process is incorporated in the concept of
the (re) laxative. The laxative is a rearranging device. It
is the means whereby one can escape from established
patterns and create new ones. The laxative allows the
arrangement of information in new ways from which
new patterns can arise. The concept of the laxative is
crystallized into a definite language tool. This language
tool is PO. Once one learns the function and use of PO

one has learned how to use lateral thinking. The whole concept of lateral thinking is concentrated in the use of this language tool. Lateral thinking could be said to be the management of PO just as logical thinking is the management of NO.

PO is to lateral thinking what NO is to logical thinking. NO is a rejection tool. PO is an insight restructuring tool. The concept of the laxative is the basis of lateral thinking just as the concept of the negative is the basis of logical thinking. Both concepts have to be crystallized into language devices. It is essential to have language devices because of the passive nature of the mechanism of mind. The language devices are themselves patterns which interact with other patterns on the self-organizing memory surface of mind to bring about certain effects. Such language devices are extremely useful in one's own thinking and for communication they are essential.

Although both NO and PO function as language tools the operations they carry out are totally different. NO is a judgment device. PO is an antijudgment device. NO works within the framework of reason. PO works outside that framework. PO may be used to produce arrangements of information that are unreasonable but they are not really unreasonable because lateral thinking functions in a different way from vertical thinking. Lateral thinking is not irrational but arational. Lateral thinking deals with the patterning of information not with the judgement of those patterns. Lateral thinking is prereason. PO is never a judgment device. PO is a construction device. PO is a patterning device. The patterning process may also involve depatterning and repatterning.

Although PO is a language tool it is at the same time an antilanguage device. Words themselves are just as much

cliché patterns as the way they are put together. PO
provides a temporary escape from the discrete and
ordered stability of language which reflects the
established patterns of a self-organizing memory
system. That is why the full function of PO is unlikely to
have evolved in the development of language. Instead
PO arises from consideration of the patterning behaviour
of the mind.

The function of PO is the rearrangement of information
to create new patterns and to restructure old ones.
These two functions are but different aspects of the
same process but for convenience they may be
separated.
● Creating new patterns.
● Challenging old patterns.
These two functions can be expressed in another way:
● Provocative and permissive: putting information
together in new ways and allowing unjustified
arrangements of information.
● Liberating: disrupting old patterns in order to
allow the imprisoned information to come together in a
new way.

The first function of PO: creating new arrangements of information

Experience arranges things in patterns. Things in the
environment may happen to be arranged in a particular
pattern or else attention may pick things out in a
certain pattern. In one case the pattern is derived from
the environment and in the other case it is derived from
the memory surface of mind since this directs attention.
The first function of PO is to create arrangements of
information that do not arise from either of these two
sources. Just as NO is used to weaken arrangements that
are based on experience so PO is used to generate
connections that have nothing to do with experience.

Once information has 'settled' into fixed patterns on the
memory surface* then new arrangements can only
occur if they are directly derived from these patterns.
Only such trial arrangements of information are
allowed as would be consistent with these background
patterns. Anything else is dismissed at once. Yet if
(somehow) different arrangements of information
could be brought about and held for a short while then
the information might snap together to form a new
pattern that was either consistent with the background
pattern or capable of altering it. This process is shown
diagrammatically opposite. The purpose of PO is then
either to bring about arrangements that would
otherwise not occur or to protect from dismissal
arrangements that would otherwise be dismissed as
impossible. These functions may be listed as follows:
To arrange information in a way which would never
have come about in the normal course of events.
To hold an arrangement of information without
    judging it.
To protect from dismissal an arrangement of
information which has already been judged as impossible.

An arrangement of information is usually judged as
soon as it comes about. The judgment results in one of
two verdicts: 'This is permissible', or 'This is not
permissible'. The arrangement is either affirmed or
denied. There is no middle course. The function of PO
is to introduce a middle course as suggested in the
diagram. PO is never a judgment. It does not quarrel with
the verdict but with the very application of the
judgment. PO is an antijudgment device.

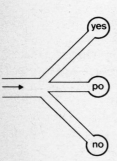

PO allows one to hold an arrangement for a little longer
without having to affirm or deny it. PO delays judgment.

The usefulness of delaying judgment is one of the most
basic principles of lateral thinking. It is also one of the

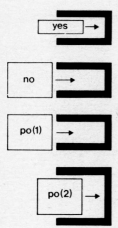

fundamental points of difference from vertical thinking.
With vertical thinking an arrangement of information
must be right at every step, which means that one must
use judgment at the earliest possible opportunity. With
lateral thinking an arrangement of information may be
wrong in itself but can lead to a perfectly valid new
idea. This possibility arises directly from consideration
of the mind as a self-maximizing memory surface.

By delaying judgment and holding onto an idea a
number of things may happen. If the idea is pursued
far enough it may be found to make sense. If one holds
onto the idea then freshly arrived information can
interact with the idea to give a valid idea. The unjudged
idea may direct the search for information that can
prove useful in its own right. Finally if the idea is held
long enough then the context into which it did not fit
may itself be changed.

Exactly the same considerations apply to the use of PO
for protection of arrangements of information that have
already been judged and dismissed. Such dismissed
arrangements may have been dismissed long ago and it
may be a matter of resurrecting them under the
protection of PO. On the other hand the arrangements
may have been proposed and dismissed only recently.

It is important to realize that the use of PO for creating
new arrangements of information is quite different from
the use of the usual devices for arranging information.

PO does not have an addition function as provided by
'and'.
PO does not have an identity function as provided by 'is'.
PO does not have an alternative function as provided by
'or'.

The function of PO *is to bring about a provocative*

arrangement of information without saying anything at all about it. The arrangement itself is not important but what happens next is. The purpose of the arrangement is to lead forward to new ideas.

In practice there are certain specific occasions on which it is convenient to use PO.

● Juxtaposition
The simplest use of PO is to hold two unrelated things together in order to allow them or their associations to interact. No connection or relationship at all is implied between the two things. Nor is there any reason for putting them together (except what might happen). Without the PO device one would not easily be able to put things together in this way without finding, suggesting, or forcing some reason.
One might say, 'computers PO omelettes'.
From this juxtaposition might come such ideas as:
Cooking by computer or by some preset automatic device. Another idea would be a central store of recipes and one would use a telephone to dial in your ingredients and requirements in order to be given a matched recipe. Both omelettes and computers are concerned with the changing of raw material into a more usable form. In an omelette things are mixed up but come out in a definite form so with a certain type of computer an apparently random mixing of information would still result in some definite output (as for instance in the brain).

● Introduction of a random word
Instead of linking two unconnected words together as in juxtaposition PO can be used to 'introduce' a random unconnected word into a discussion in order to stimulate new ideas. You could say, 'Gentlemen you know all about lateral thinking and the use of a random input to help disturb cliché patterns of thought and to

stimulate new ideas. I am now going to introduce such a
random word. This word has no connection at all with
what we have been discussing. There is no reason
behind my choice of the word. The only reason for its
use is the hope that it will provoke some new ideas. Do
not feel that there really is a hidden reason. Do not
spend your time searching for this reason. The word is
'raisin'. Instead of saying all that one would simply say:
'Po raisin'.

If the problem under discussion was, 'How to use study
time' then this random word could set off such ideas as:
raisin – used to make cakes enjoyable – small pockets of
sweetness – intersperse short periods of more interesting
subjects among longer periods of less interesting subjects
– create small nodes of interest in less interesting
subjects.
raisins – dried grapes – concentrated sweetness –
concentrate and summarize material so that it can be
taken in over a shorter time.
raisins – exposed in the sun to dry – perhaps one can
study in a pleasant surrounding as easily as in an
unpleasant one – does lighting, colour etc affect
boredom. Perhaps material can be subjected to 'glare' of
analysis by someone else in order to reduce it to its
essentials.
raisins – dried for preservation – notes and summaries
easier to remember but need reconstituting with fluid
(i.e. examples).

● Disconnected jumps
In vertical thinking one moves in sequential steps but in
lateral thinking one can make disconnected jumps and
then try and fill in the gaps. If you do this in the middle
of a vertical thinking discussion then everyone else will
be very confused as they try to find the logic behind
this jump. In order to indicate that the jump is a lateral
disconnected one you could preface your comment

with PO. For instance in the discussion about study time
you might say, 'Po time spent studying is time spent not
doing other things'.

The jump may be only a small one within the same field
or it may be a large one to an unconnected field. PO
saves one the trouble of having to link the new remark
to what has gone before. As usual PO implies, 'Don't
look for the reason behind this. Let us just go forward
and see what the *effect* of it is.'

● Doubt (semi-certainty)
Whenever a discussion gets blocked by the impossibility
of proving a certain point PO can be used to open things
up again. PO does not prove the point or deny it but it
allows the point to be used in any way which will enable
the discussion to keep going. One can then see what
happens. It may be that nothing very useful comes of it
and one realizes that the original point was not so vital
after all. It may be that one can reach a solution and
from this one can find another way back to the starting
point without having to go through the doubtful point.
It may be that one can only reach a solution through the
doubtful point and so one comes to realize how vital this
point is and therefore increases the effort to prove it.
This particular use of PO is not very different from the
ordinary use of 'if' or 'suppose'.

● Being wrong
In lateral thinking one does not mind being wrong on
the way to a solution because it may be necessary to go
through a wrong area in order to get to a position from
which the correct path is visible. PO is an escort that
allows one to move through the wrong area. PO does not
make things right but it switches attention from why
something is wrong to how it may be useful. In effect
PO implies, 'I know this is wrong but I am going to put
things this way in order to see where it leads me'.

In considering the problem of keeping the windscreen of a car free from dirt and water someone suggested that cars ought to be driven backward since the back window was always much easier to see out of than the front window. In itself this is obviously nonsense since if one was going backwards that window would get just as dirty as the ordinary windscreen. Nevertheless the suggestion, 'Why not drive backwards' can lead on to such other ideas as indirect vision systems or some way of protecting the windscreen from head on exposure to mud and water.

In this example PO would be used in the following way. Someone would suggest driving backwards and this would be met with the response, 'That's nonsense, because . . . . ' The reply to this would be, 'Po why not drive backwards'. The purpose of PO would be to delay judgment – to hold the idea in mind for a few moments in order to see what could arise from it instead of dismissing it at once.

● Holding function
In addition to protecting an idea which is obviously wrong PO can be used to protect an idea from judgment. In this case the idea has not already been judged but is about to be subjected to critical analysis. PO is used to *delay* this. This function of PO is rather similar to its use for the introduction of a random stimulus. An ordinary remark or idea in the course of a discussion is turned by the use of PO into a catalyst. Used in such circumstances PO indicates: 'Let's not bother to analyse whether this is right or wrong – let us just see what ideas it will lead to.'

PO could be used by the person offering the idea or it could be used by anyone else. Thus if an evaluation of the idea was started someone could simply interject, 'Po . . . . ' This would mean, 'Let's hold off evaluation for the moment'.

● Construction
In school geometry a problem is often made easier to
solve by adding some additional lines to the original
figure. This process is similar to that involved in the
story of the lawyer whose task it was to divide up eleven
horses among three sons so that one of them got half of
the horses, another got a quarter, and the third son got
one sixth. What he did was to lend his own horse to the
sons and then divided the twelve horses up, giving the
first son six, the second three and the third two. He then
took his own horse back again.

Here PO is used to add something to the problem or to
change it in some other way. Changing the problem in
this way can lead to new lines of development, new
ways of looking at it. The purpose of changing the
problem is *not* to rephrase it or put it in a better way but
to alter it and see what happens next. For instance in
considering the efficiency of the police in dealing with
crime one might say, 'Po why not employ one-armed
policemen?' Changing the problem in this way by
adding the factor of 'one-armed policemen' would focus
attention on the possible advantages of being one-armed
and especially on the need to use brain and organization
rather than muscle power.

Summary
There are many other ways in which PO can be used but
the occasions listed above are enough to illustrate the
first function of PO. This first function is quite simply to
allow one to say anything one likes. PO allows one to
arrange information in any way whatsoever. There need
be no justification at all for such arrangement except PO.

Po two and two make five.
Po water flows uphill if it is coloured green.
Po lateral thinking is a waste of time.
Po men have souls and women have not.

Po it takes a lifetime to unlearn what has been learned in education.

The first function of PO is to shift attention from the meaning of a statement and the reason for making it to the effect of the statement. With PO one looks forwards instead of backward. Because any arrangement of information can lead on to other arrangements a statement can be very useful as a stimulus no matter how nonsensical it is in itself. And by being nonsensical one can arrange information in a way that is different from the established patterns – and so increase the chance of a permanent restructuring. With vertical thinking one is not allowed to do any of this. With vertical thinking one looks backwards at the reason for a statement, at the justification, at the meaning.

The statement, 'Po water flows uphill if it is coloured green' is ridiculous but it could lead on to such ideas as: Why should the green colour make a difference? Why should adding colour make a difference? Is there anything one could add to water to make it flow uphill? In fact there is. If one adds a very small amount of a special plastic then the water acts as a solid/liquid to such an extent that if you start pouring water out of a jug and then hold the jug upright the water will continue to siphon out, climbing up the side of the jug, flowing over the rim and down the outer side.

PO as a device allows one to use information in this way which is completely different from the ordinary use of information. One could use information in this way without PO but one would still be using the lateral concept which is incorporated in PO. The convenience of PO as an actual language device is that it clearly indicates that information is being used in this special way. Without such an indication there would be confusion as the listener would not know what was

going on. A PO type statement inserted into an ordinary
vertical thinking discussion without the use of PO would
lead the listeners to suppose that the speaker was mad,
lying, mistaken, stupid, ignorant or facetious. Apart
from the inconvenience of being the recipient of such
judgments there is the danger of being taken seriously.
For instance, 'Po the house is on fire' is rather different
from 'The house is on fire'. Furthermore if one does not
use PO then the information is not used as a stimulus in
the lateral manner.

The second function of PO: challenging old arrangements of information

The basic function of mind is to create patterns. The
memory surface of mind organizes information into
patterns. Or rather it allows information to organize
itself into patterns.* The effect is just the same as if the
mind picked things out of the environment and put
them together to give patterns. Once formed these
patterns become ever more firmly established because
they direct attention. The effectiveness of mind depends
entirely on the creation, the recognition and the use of
patterns. The patterns have to be permanent to be of
any use. Yet the patterns are not necessarily the only
way of putting together the information contained in
them – or even the best. The patterns are determined by
the time of arrival of the information or by preceding
patterns that have been accepted entire.

The second function of PO is to challenge these
established patterns. PO is used as a freeing device to
free one from the fixity of established ideas, labels,
divisions, categories and classifications. The way PO is
used can be summarized under the following headings:

● To challenge the arrogance of established patterns.
● To question the validity of established patterns.
● To disrupt established patterns and liberate

information that can come together to give new patterns.
- To rescue information trapped by the pigeonholes of labels and classifications.
- To encourage the search for alternative arrangements of the information.

- Never a judgment
As suggested before PO is never used as a judgment device. PO is never used to indicate whether an arrangement of information is right or wrong. PO is never used to indicate whether an arrangement of information is likely or unlikely or whether it is the best available at the moment. PO is a device to bring about an arrangement or rearrangement of information not a device to judge the new arrangements or condemn the old ones.

PO implies, 'That may be the best way of looking at things or putting the information together. That may even turn out to be the only way. But let us look around for other ways.'

With vertical thinking one is not allowed to challenge an idea unless one can show why it is wrong or else provide an alternative. If one provides an alternative one must somehow show why this alternative is preferable to the original idea as well as proving that the alternative is sound. With PO one has to do none of these things. One challenges the established order without necessarily being able to offer anything in its place or even to show any deficiency.

Judgment usually asks for justification of an idea. Justification of why an arrangement of information should be accepted. One wants to know why something has been put together in a certain way. With PO the emphasis is shifted away from this 'why' to 'where to'. One accepts the need to rearrange information in new

ways. One takes a new arrangement and instead of
trying to see where it has come from and whether it is
justified one sees where it leads to—*what effect it can
have.*

● The response to PO
The challenge of PO is not met by a fierce defence of
why the established idea is indeed the best possible
way of putting things together because PO does not
attack an idea. PO is a challenge to try and think of
other ways. The challenge of PO is met by generating
different ways of looking at the situation. The more
ways one can generate the more clearly it may be shown
that the original idea was indeed the best one but that
is no reason for refusing to try and generate other ways.
If in generating these alternative ways a new and
better way of looking at things turns up then that can
only be a good thing. Even if the old idea is only
altered slightly that is still a good thing. Even the
possibility that there might be another way of looking
at things is useful in itself in so far as it lessens the
rigidity of the old idea and makes it more easily changed
when change is due.

● Challenging cliché patterns
Any pattern that is at all useful is a cliché. The more
useful it is the more of a cliché it tends to become.
And the more of a cliché it is the more useful it may
become. PO can be used to challenge any cliché. PO
not only challenges the way concepts are arranged into
patterns but the very concepts themselves. One always
tends to think of clichés as arrangements of concepts
but that the concepts themselves must be accepted as
the building blocks of thought and so must themselves
remain unaltered.

'Po freedom' challenges the very concept of freedom
not the value or purpose of freedom.

'Po punishment' challenges the very concept of punishment not the circumstances under which it is used or the purpose for which it is used.

As suggested above it is the useful concepts that need challenging most. The less useful concepts are likely to be under perpetual challenge and reformation. But the usefulness of a useful concept protects it.

● Focusing
Since the cliché may refer to a particular concept or a phrase or to the whole idea it is helpful if one is specific about what is being challenged by PO. In order to do this one would repeat what is being challenged but preface it with PO.

'It is the function of education to train the mind and to pass on to it the knowledge of ages.'

To this one might reply: 'Po, train the mind' or 'Po the knowledge of ages', or even just 'Po train'.

Used in this way PO can act as a *focusing* device to direct attention to some concept that is always taken for granted because there are other concepts which seem more open to reexamination.

● Alternatives
There are times when it is reasonable to try and find other ways of looking at a situation. This happens when the current approach is not satisfactory. PO is used as a demand to generate alternatives even when it is quite *unreasonable*. One goes on generating alternatives right up to the point of absurdity – and beyond. Since there is no good reason for generating alternatives under these circumstances one needs the artificial stimulus of PO which is a device that works outside of reason.
'It is spring and the bird is on the wing.'

'No. The wing is on the bird.'
'Po'
'The bird and the wing both happen to be going along
in the same direction.'

Used in this way PO is an invitation (or a demand) to
generate alternative arrangements of the information.
It is also used to justify those alternative arrangements
by making it clear that they are offered as alternative
arrangements and not necessarily better arrangements
or even justified ones.

● Antiarrogance
One of the most valuable functions of PO is as an
antiarrogance device. PO is a reminder of the behaviour
of the memory surface of mind. PO is a reminder that a
particular arrangement of information which seems
inevitable may yet have come about in an arbitrary
fashion. PO is a reminder that the illusion of certainty
may be useful but that it cannot be absolute. PO is a
reminder that certainty about a particular arrangement
of information can never exclude the possibility of
there being another arrangement. PO challenges
dogmatism and absolutism. PO challenges the
arrogance of any absolute statement or judgment or
point of view.

Used in this way PO does not imply that the statement
is wrong. It does not even imply that the person using
PO has doubts about the statement let alone justified
doubts. All PO implies is that the statement is being
made with a degree of arrogance that is not justified
under any circumstances.

PO implies the following: 'You may be right and your
logic may be faultless. Nevertheless you are starting
from perceptions that are arbitrary and you are using
concepts that are arbitrary since both are derived from

your own individual experience or the general experience of a particular "culture". There are also the limitations of the information processing system of mind. You may be right within a particular context or using particular concepts but these are not absolute.'

PO used in this fashion is never intended to introduce so much doubt that an idea becomes unusable. PO is never directed at an idea itself but only at the arrogance surrounding it – at the exclusion of other possibilities.

● Counteracting NO

NO is a very convenient device for handling information. It is a very definite and a very absolute device. NO also tends to be a permanent label. The permanence of the label, its definiteness and its absolute rejection, may rest on evidence that was at best flimsy. Once the label is applied however then the full force of the label takes over and the bare adequacy of the reason behind its application is lost. It may also happen that the label was justified when it was originally applied but that things have changed and the label is now no longer justified. Unfortunately the label remains until it is removed – it does not only last so long as there are reasons for it to last. Nor is it easy to examine whether there are sufficient reasons for maintaining the label because one cannot know whether a label is worth reexamining until one has in fact done so and the NO label itself deters such examination.

PO is used to counteract the absolute block caused by the NO label. As usual PO is not a judgment. PO does not imply that the NO label is incorrect nor does it even suggest that there is doubt about the label. In effect PO implies: 'Let us cover up that NO label for the moment and proceed as if it was not there.' As one goes forward with one's examination it may become obvious that the

label is no longer justified. On the other hand it may
become obvious that the label is still as valid as ever but
nevertheless information which has been hidden behind
the label may be very useful elsewhere.

Consider the statement: 'You cannot live if your heart
stops.' This would be changed to 'Po you can live if
your heart stops' and this leads on to consideration of
the artificial devices for keeping a heart beating, for
artificial hearts or transplanted hearts. It also leads on
to the need for a new criterion of death since the heart
can be kept beating by artificial means even when the
brain is irreversibly damaged.

The history of science is full of instances when
something was said not to be possible but later proved
to be possible. Heavier than air flying machines are an
example. In 1941 someone showed that to get a load
weighing one pound to the moon would require a
rocket weighing one million tons. Eventually the
rocket that actually sent men to the moon weighed far
less.

Any definite use of the NO label is an invitation to use
PO.

● Antidivision
In so far as PO is used to challenge concepts it also
challenges the division which divides something into
two separate concepts. PO challenges not only the
concepts but the division that has brought them about.
The pattern making tendency of mind can both put
together things that ought to be separated and also
separate things that ought to be put together. Both an
artificial difference and an artificial sameness may be
challenged with PO.

If two things are separated by a division then PO may

challenge the division or may shift attention towards the
features which the two things have in common and
away from those features that separate them.

Rigid divisions, classifications, categories and
polarizations all have a great usefulness but they can
also be limiting. As with NO the function of PO is to
temporarily lift the labels and let the information come
together again for reassessment. Information is dragged
out of pigeonholes and allowed to interact. Things may
be classified by a particular feature or by a particular
function. Once classified the label becomes permanent
and as a result all the other features and functions tend
to be forgotten. One does not think of looking under a
label for a function that is not indicated on that label.
As in a filing system something is more effectively lost
if it is misfiled than if it is not filed at all.

A spade and a broom are two very different things.
'Spade po broom' focuses attention on the similarities:
in both a function is performed at the end of a shaft,
both have long shafts, both can be used in a
right-handed or a left-handed manner, in both there is
a wide part at the end of the narrow part, both can be
used for removing material from a place, both could
be used as a weapon, both could be used to prop a door
open etc.

'Artist po technologist.' One is very ready to put
people into pigeonholes and the further the pigeonholes
are apart the more useful they seem to be. They seem to
be more useful because with far apart pigeonholes one
finds it easier to predict what a person is going to do
than if the pigeonholes overlapped. 'Artist po
technologist' challenges the big gap there is supposed to
be between the two types. It suggests that the two types
may be both trying to do the same thing: to achieve an
effect. The materials may be different but the methods

may be the same: a combination of experience, information, experimentation and judgment. It may also suggest that nowadays an artist has to be something of a technologist if he is to use the newer media.

● Diversion
PO challenges concepts, it challenges the division between concepts, and it can also be used to challenge the line of development of a concept. Sometimes the line of development of an idea is so natural and so obvious that one moves quite smoothly along this path before ever wondering whether there might be an alternative path to be explored. To prevent this PO may be used as a temporary blocking device. PO is used as a special sort of NO but without the judgment of NO or the permanence of NO. In effect PO implies: ''That is the natural path of development but we are going to block that path for the moment in order to make it possible to explore some other pathways.'

'A business exists to make profits. Profits are obtained from the most efficient methods of production coupled with thorough marketing and the maximum price the market will bear . . . .' This is a natural and reasonable line of thought. But if one were to challenge, 'Po to make profits' then one would be able to explore other possible developments. 'A business has the social function of providing an environment in which people can make the maximum contribution to society through productivity.'
'A business exists as an efficient production unit. Efficiency is the main aim not profit.'
'A business only exists as an evolutionary stage in the organization of production and its only justification is historical.'

If PO is used skillfully it can divert the line of thought into new pathways by blocking the old ones at certain

crucial points. PO is an excuse for choosing a line of
thought that is not the most obvious or the best.

● PO and overreaction
The general function of PO is as a laxative to relax the
rigidity of a particular way of looking at things. In
certain situations a rigid way of looking at things can
lead to emotional overreaction. In such cases PO acts as a
laugh or a smile to release the tension that accompanies
a rigid point of view. Both a laugh and a smile occur
when a particular way of looking at a situation is
suddenly turned around. PO suggests the possibility of
such a change in view. PO acts to lessen the fierce
necessity of a particular point of view.

General function of PO
PO is the laxative of language and thinking. PO is the
device for carrying out lateral thinking.

PO is a symbol which draws attention to the pattern
making behaviour of mind which tends to establish
rigid patterns. PO draws attention to the possibility of
clichés and rigid ways of looking at things. PO draws
attention to the possibility of insight restructuring to
obtain new patterns without any further information.
Even if PO is never used except as a reminder of these
things then it can still be extremely useful.

When used as a practical language tool the function of
PO is to indicate that lateral thinking is being used. PO
indicates that the arrangement of information being
made makes sense from a lateral thinking point of view
even if it does not make sense otherwise. Without some
definite indicator such as PO there would be confusion
when lateral thinking was introduced in the middle of
an ordinary vertical thinking discussion.

PO is not a selective device but a generative one. PO is

never a judgment. PO never examines why an arrangement of information has been made but looks forward to what effect it may have. PO does not oppose or counteract judgments but merely sidesteps them. PO also protects arrangements of information from judgment.

PO is essentially a device to enable one to use information in a way that is other than the most obvious and the most reasonable. PO allows one to make arrangements of information for which there is no justification. PO also allows one to challenge arrangements of information for which there is full justification.

PO may seem a perversion designed to upset the highly useful system of logical thinking, permanent concepts and the pursuit of the most obvious. PO is not however a perversion but an escape. It does not destroy the usefulness of this system but adds to it by overcoming the rigidity which is the main limitation of the system. It is a *holiday* from the usual conventions of logic not an attack upon them. Without the stabilizing background of traditional vertical thinking PO would not be much use. If everything was chaos then there would be no rigidity to escape from nor would there be any possibility of establishing a more up to date pattern which is what insight is about. As a device PO actually enhances the effectiveness of vertical thinking by keeping it intact. This PO does by providing a means to bypass vertical thinking in order to introduce a generative factor. Once a new pattern has emerged it can be developed with the full rigour of vertical thinking and judged.

Similarity of PO to other words
It may be felt that some of the functions of PO are very similar to those carried out by such words as hypothesis, possible, suppose and poetry. There are some functions

of PO which are indeed similar for instance the
semicertainty function. But there are other functions of
PO which are quite different for instance the
juxtaposition of totally unrelated material. Hypothesis,
possible and suppose are very weak relations of PO.
They cover arrangements of information which seem
very reasonable but cannot quite be proved. They are
tolerable guesses at the best arrangement of information
at the moment. PO in contrast allows information to be
used in ways which are *totally* unreasonable. The most
important difference is that with these words the
information is used for its own sake even if the use is
tentative. With PO however the information is not used
for its own sake but for its effect. Perhaps the most
similar word is poetry where words are used not so
much for their own meaning as for their stimulating
effect.

The mechanism of PO

Why should PO work? PO could never work in a linear
system like a computer because the arrangement of
information in such a system is always the best possible
one according to the programme. But in a
self-maximizing system or a system with humour the
arrangement of information into patterns depends very
heavily on the sequence of arrival of information. Thus
A followed by B, followed by C, followed by D, would
give a different pattern to B followed by D, followed by
A, followed by C. But if A, B, C and D were all to arrive
together then the best arrangement of them would be
different from either of the other two arrangements.
There is a tremendous continuity in this type of system
and this means that it is easy to add to patterns or
combine them but very difficult to restructure them.*
There are also the inherited patterns which are
acquired ready made from other minds.

Because of this tendency to establish patterns and for

them to become ever more rigid one needs a means for disrupting the patterns in order to let the information come together in new ways. PO is that means as it is the tool of lateral thinking. PO is needed because of the behaviour of a self-maximizing memory system and PO works because of the nature of such a system. Within such a system some sort of pattern has to form. If the old pattern is sufficiently dislocated then a new pattern is formed and the process is insight restructuring.

PO is used to disrupt patterns. PO is used to dislocate patterns. PO is used as a catalyst to bring together information in a certain way. From that point on it is the natural behaviour of the mind that snaps the new pattern together. Without such behaviour PO would be useless.

The bigger the change from the old pattern the more likely is a new pattern to snap together. 'Reasonable' arrangements of information are too closely similar to the old arrangements to give new patterns. That is why PO works outside of reason. PO is concerned not with the reason for using information in a certain way but for the effect it will have. Once the new pattern has come about it must of course be judged in the usual way.

In emptying a bucket by a siphon the water must first be sucked upwards in the tube. This is an unnatural direction for water to travel. Once the water has reached a certain position then the siphon forms and the water will continue to flow naturally out of the bucket until it is empty. In the same way an unnatural use of information may be necessary to provoke a rearrangement that is itself perfectly natural.

Grammatical use of PO
PO can be used in any way that seems natural. The most important point is that anything covered by PO should

be clearly seen to be covered by PO. The two main
functions of PO are first to protect an arrangement of
information from judgment and to indicate that it is
being used provocatively and second to challenge a
particular arrangement of information such as an idea, a
concept or a way of putting things. In the second case
the material being challenged would be repeated and
PO would be added to it. In the other case PO would
cover new material.

1   PO as interjection
Here PO would be used by itself as a reply or even as an
interruption much as NO is used. It would imply that a
particular way of looking at things was being
challenged.
e.g. 'The purpose of sport is to encourage the
competitive spirit and the will to win.'
'Po!'

2   PO as preface
Here PO is used before a sentence or a phrase or a
word that it is meant to qualify. The qualification may
take the form of a challenge or it may take the form of
introducing provocative material.
e.g. 'An organization can only function efficiently if all
its members show absolute obedience.'
'Po function efficiently.'
or 'Po clockwork with the cogwheels made of rubber.'

3   PO as juxtaposition
When two words are going to be juxtaposed for no
reason at all PO is used to indicate this relationship
between them. This same use of PO is involved in the
introduction of a random word into a discussion.
e.g. 'Travel po ink.'
or 'Po kangaroos.'

4   PO in the same positions as NO or NOT

PO can be used in any position in which NO or NOT could be used. In such a position PO would qualify exactly the same things as NO or NOT would qualify. e.g. 'Wednesday is po a holiday.'

In practice it is probably best to try to use PO always at the beginning of a sentence or phrase or right in front of the word to be qualified. PO does not have to be written in capital letters but until one is well used to it capital letters are preferable. If one is using PO and the other person does not understand its use then this can be most simply explained as follows:

1   Challenge function
PO means you may very well be right but let's try and look at it in another way.

2   Provocative function
PO means I am just saying that to see what it sets off in your mind, to see whether that way of putting things can stimulate any new ideas.

3   Antiarrogance function
PO means don't be so arrogant, so dogmatic. Don't have such a closed mind.

4   Overreaction
PO simply means, let's cool it. There is no point in getting upset about this.

Practice

PO is the language tool of lateral thinking. The concept and function of lateral thinking is crystallized in the use of PO. If one acquires skill in the use of PO then one has skill in the use of lateral thinking. For this reason practice in the use of PO is extremely important. Learning how to use PO is similar to learning how to use NO. Learning how to use NO is however a gradual process spread over many years. With PO one tries to achieve the same effect in a shorter time. It is much better to go slowly and carefully than to rush ahead and teach only a limited or even incorrect use of PO.

In teaching the use of PO it is far better to suggest the general concept of PO than to define rigidly the situations in which it can be used. Nevertheless one needs to show the practical use of PO in language and not just the theory behind it.

Since PO is the tool of lateral thinking any of the previous practice sessions could be reused with PO as the operative device. It is more useful however to devise special situations which indicate the function of PO more specifically.

In this section several aspects of the function of PO have been listed. These aspects can be mentioned in the course of explaining the nature of PO and as one mentions them one can give and ask for further examples. For the actual practice session it is better to group the functions of PO into a few broad uses than to confuse with the detail of each particular use.

The function of PO involves two basic aspects:
- The use of PO.
- The response to PO.

The response to PO
It is far better to learn the response to PO *before* the use of PO. The reason for this apparently paradoxical arrangement is that by learning how to respond to PO one actually learns the reason for using it. In addition by learning the response first one can then practise the use of PO in a more realistic way since it will not only be used but also responded to.

The points about the response to PO are as follows:

1   PO is never a judgment. This means that when PO is used to challenge something that you have said this does not imply disagreement or even doubt. PO is *never met* with a defence of what has been said. Nor is PO met with an exasperated, 'How else could it be put – how would *you* put it?' Furthermore PO is not an indication that the person saying it has a better alternative or even an alternative at all. What PO implies is, 'Without disagreeing with what you say let us – both of us – try and put things together in a different way. It is not me against you but a joint search for an alternative structuring.' It is important to stress this aspect of the *joint search*. It is important to stress that PO is not part of the antagonism of an argument. So one responds to PO by trying to generate alternatives *not* by irritation or by defending the original way of putting things.

2   PO may involve the provocative use of information. This means that information may be put together in a fantastic and completely unjustified way which is covered by PO. In responding to this use of PO one does not argue that the arrangement of information is unacceptable. One does not demand the reason for putting things together in this way. Nor does one sit back and imply, 'Very well if you want to put things like that you go ahead and show that it can be useful.' The provocative use of PO is to provide a stimulus which is to be used cooperatively by both parties. It implies: 'If we use this arrangement of information as a stimulus what can we both come up with?' So the response to the provocative use of PO is neither condemnation nor indifference but active cooperation.

3   PO may be used as a protection. This means that PO may be used to hold off judgment or to temporarily override a judgment that has resulted in a rejection. The response to this use of PO is *not to show that the judgment*

is necessary and should be applied at once. Nor is the response one of exasperation, 'If you won't accept the ordinary uses of right and wrong how can we ever proceed?' Nor is the response one of superior indifference, 'If you want to say that black is white and to play around with that idea for a while I shall just wait until you are through.' As before the proper response is a cooperative exploration of the new situation.

4   PO may be a relaxation. This means that when a situation has become tense through the development of rigid points of view and possibly overreactions, PO is suggested as a smile to relax the tension and to relax the rigid points of view. Here the only appropriate response is to respond with PO (with a mental shrug and a smile) and to relax the rigidity of the situation.

5   PO may be used ambiguously. There are times when it is not clear how PO is being used or what concept is being challenged. In such cases one simply asks for the person using PO either to be more specific or to agree that he really wants to use it in a general way.

In summary one may say that the most important aspect of the response to PO is to realize that it is not directed against anything but is a suggestion for cooperative attempts to restructure a situation. If one feels competitive then one can express this by using PO more effectively than the person suggesting it: that is to say one goes on to generate more alternatives than he does. PO may be an invitation to a race but never an invitation to a conflict.

The use of PO
For convenience the many uses of PO may be divided into three broad classes.

1   The generation of alternatives. Antiarrogance.

Relaxation. Reexamination of a concept. Rethinking. Restructuring. Indicating an awareness of the possibility of clichés or a rigid point of view.

2   Provocation. The use of arrangements of information as stimuli. Juxtapositions. Introduction of random words. Abolition of concept divisions. The use of fantasy and nonsense.

3   Protection and rescue. Holding off judgment. Temporarily reversing judgment. Removal of the NO label.

The generation of alternatives
PO is used to point out that a particular way of looking at a situation is only one view among many. PO is used to point out that a particular point of view appears to be held with an unjustified arrogance. The first level is merely to suggest that there may be other ways of looking at the situation. This is especially so when one uses PO as an antiarrogance device.

The next level is to invite restructuring of the situation. Here one asks for alternatives and goes on to supply them oneself.

PO may be applied to a whole idea, a whole sentence, a phrase, a concept or just a word.

Practice
1   The teacher asks a student (a particular student or a volunteer) to talk on some subject. The subject could be something like the following:
What is the use of space travel?
Should all medical aid be free?
Are straight roads better than winding ones?

In the course of the student's talk the teacher interrupts

with PO. The interruption repeats part of what the student has said and prefaces it with PO. The student is not expected to respond to PO at this stage. This is explained to him. He just pauses while the teacher interrupts and then carries on.

2   The teacher talks about a subject and this time the students are invited to interrupt with PO in the same way as the teacher had done in the preceding practice session. Subjects for discussion might include:
The usefulness of different languages.
Whether large organizations work better than small ones.
Was it easier to work alone or in a group?

Each time a student interrupts with PO the teacher responds by generating alternative ways of putting things and the students are encouraged to do the same. For example a discussion might go something like this:

TEACHER: Different languages are useful because they allow the development of different cultures and so provide more interest.
STUDENT: PO provide more interest.
TEACHER: Different cultures mean different ways of looking at life, different habits and ways of behaving, different art etc. All these are things one can learn about and find out about and compare to one's own. New patterns to be explored. Something to be done.
STUDENT: Different ways of expressing the same thing– they could be useful, they could be a waste of time.
TEACHER: Because of the different language communication is poor and so distinctness emerges instead of a general uniformity.
STUDENT: PO communication is poor.
TEACHER: People cannot talk easily to people with another language or read their books. People cannot influence each other so much.

STUDENT: People cannot influence each other. That may be a bad thing because from such interaction might come better understanding.
TEACHER: PO understanding.
STUDENT: They would know what the other person meant, what he was up to, what he wanted, what his values were.

3   It is quite likely that a discussion of this sort would very quickly become a two way discussion. If not then the teacher can deliberately arrange for a debate type discussion between two students. Each of them is allowed to use PO and so is the teacher who can interrupt with PO but is not allowed to take part in the discussion otherwise.

Comment
In this type of discussion it may become obvious that PO is being used mainly as a focusing device to indicate: 'explain what you mean by . . .' or, 'define that . . .' or, 'elaborate that point . . . .' If this seems to be the case then the teacher points out that the function of PO is to ask for a restructuring, to ask for *alternative* ways of putting things. When PO is next used the teacher calls for a pause and then invites the entire class to list different ways of putting whatever has been qualified by PO. For instance 'Po understanding' from the example given above might rise to the following:
Supposing that the other person reacts in the same way as you.
Things mean the same to the other person as to you.
Lessen the possibility of misunderstanding.
Full sympathy.
Communication without interpreters or intermediaries.
Ability to listen and respond.

None of these are complete or even very good definitions of 'understanding' but they are different ways of putting things. Perhaps the best of them is 'lessen the chance of misunderstanding'. This may seem a tautology but from an information point of view it says a great deal.

4   Picture interpretation. This is similar to the picture interpretation that was practised in an earlier session. The caption is removed from a photograph and a student (or students if there are enough copies of the photograph or other means for making it visible to all) is asked to interpret it. He offers an interpretation and then the teacher replies, 'Po'. This simply means, 'Very well. Go on. Generate another alternative. What else could it mean?'

This is a very simple use of PO but it is helpful to practise it since it indicates the use of PO in a much clearer manner than the other situations.

Provocation
This second use of PO simply indicates that the arrangement of information has no justification except the possibility that it might set off new lines of thought. Such an arrangement of information may be as fantastic or unreasonable as anyone can make it. The arrangement is not examined in itself but only in terms of what it sets off.

5   Juxtaposition. This is the simplest provocative arrangement of information. Two words are put together with PO inserted between them to indicate why they are put together. The pairs of words are then offered to the class one at a time. The session may be conducted in an open class with students volunteering suggestions which are listed by the teacher on a blackboard or else by some student who is asked to take notes. Alternatively the students can list their own ideas and these are collected and compared at the end.

Possible pairs of words might include:
Ice cream po electric light.
Horse po caterpillar.
Book po policeman.
Rain po Wednesday.
Stars po football.
Stars po decision.
Shoe po food.

The students are not specifically asked to relate the
words, or to find some connection between the two or to
show what the two words have in common. Any sort of
ideas at all that arise are accepted. There is no question
of directing the sort of ideas that the students ought to
be having. If on reading through the results one cannot
see the connection then one asks how it came about, one
asks for the missing links. One does not care what the
idea is but one does want to know how it came about.

6   Random word. This technique has been discussed
in a previous chapter. It consists of introducing into the
consideration of a subject a word which has no
connection with the subject at all. The idea is to see
what the random word triggers off. In this case PO
would be used to introduce the random word. An
alternative way of doing it would be to take some word
which appeared to be vital in the discussion and couple
it in juxtaposition with a random word by means of PO.
Possible subjects for discussion might include:
Advantages of saving against spending.
Advantages of attack rather than defence in sport.
Knowing where to find information.
Why do fights start?
Should people do exactly what they want to?
The design of shoes.
Possible random words might include:
Fishing line.
Bus ticket.

Motor car horn.
Eggcup.

7   Concept reuniting. PO can be used to put together
again things that have been divided up into separate
concepts. PO can be used to remove labels and extract
information from pigeonholes. In order to put across
this function of PO one takes concepts which have been
created by a division (or which have created each other
by implication) and puts them together by means of PO.
Such paired concepts are presented to the class in the
same way as the juxtapositions were presented and the
ideas arising from this presentation are examined and
compared. In this instance it is better if the students
individually list their ideas so that when these are read
out at the end they can appreciate the usefulness of the
procedure.
Possible examples might include :
Soldiers po civilians.
Flexible po rigid.
Attacker po defender.
Order po chaos.
Liquid po solid.
Teacher po student.
Up po down.
Day po night.
North po south.
Right po wrong.
Male po female.

8   In addition to reacting to the juxtapositions and
paired concepts provided for them the students can be
asked to generate their own juxtapositions and paired
concepts. Suggestions for these are collected on slips of
paper and then a selection of these is fed back to the
students for their reaction. The simple exercise of
generating such juxtapositions and paired concepts *is
itself* very useful in making clear this particular use of
PO.

Protection and rescue
This function of PO is used to delay judgment. In effect it is used to delay rejection for that is the only sort of judgment which would remove an idea from consideration. PO may be used to protect an idea before it has been judged or it may be used to bring back into consideration an idea which has already been judged and rejected. In practice PO is attracted by the NO label. Whenever the NO label is used it is a direct indication of the current frame of reference against which every judgment must be made. By temporarily overriding the rejection with PO one is really reexamining the frame of reference itself.

9   A discussion is started between two students or between the teacher and a student. The discussion continues until either one or the other uses a NO rejection. At that point PO is used to overcome the rejection and the rejected statement is considered in itself to see what ideas it can trigger off.
Possible subjects for discussion might include:
Should people be encouraged to live in the country or in towns?
Does a welfare state encourage people to be lazy?
Is changing fashion in clothes a good thing?
How much should one do for oneself and how much should one pay other people to do for one?
Are classroom lessons too long?

A discussion might go something as follows:
TEACHER: People should be encouraged to live in the country because towns are not healthy.
STUDENT: Towns are not healthy. PO towns are healthy. Towns could be healthy with better planning and better traffic control. Perhaps towns could be more healthy mentally because of more social interaction.
TEACHER: Towns would have better health services because they would be more centralized and

communication would be better.

10   A subject is selected and the students are asked to
think of all the negative things they can say about that
subject. These are listed and then some of them are
reexamined using PO. Quite obviously the number of
negative things one can say about something is
infinite. For instance about an apple one could say :
'It is not black. It is not purple. It is not mauve etc. It
is not an orange. It is not a tomato etc.' In practice one
would simply ignore that sort of list or pick out of it
certain items. For instance 'An apple is not a tomato'
could lead to the following idea: 'In some languages the
word for tomato is derived from that for apple. In
Italian a tomato is called a golden apple. In Sweden the
word for an orange is derived from the word for an
apple.' To avoid this sort of thing it is probably better to
deal with abstract concepts or with functions rather
than objects.

Possible subjects might include:
Work.
Freedom.
Duty.
Truth.
Obedience.
Boredom.

General comment on the use of PO
After the initial practice sessions in which the use of PO
is obviously excessive and artificial one moves on to the
more natural use of PO in ordinary discussion sessions.
It is up to the teacher to use PO now and again to
indicate how it should be used. The other important
point is to watch how the students react to PO when it is
used either by other students or by the teacher himself.
An inappropriate reaction to PO indicates that the
function of PO has not been understood. It is more
important to emphasize the correct *reaction* to PO than

the correct *use* of it. Someone who knows how to react appropriately to PO will also know how to use it appropriately.

The one sided use of PO

PO is a device for use in one's own thinking and reacting as well as in communication with other people. In fact it is probably of more use in enabling one to use lateral thinking oneself than in allowing the use of lateral thinking in group discussions. This *private* use of PO obviously does not depend on other people understanding its function. In communication however it may come about that one person uses PO and the other person has no idea what it means. In that case one does not desist from using PO but explains what it means. Simple ways of explaining what PO means have been described earlier in this chapter. If in difficulty one could always say that it was a special form of 'suppose'.

Summary
PO is a language device with which to carry out lateral thinking. PO is an insight tool since it enables one to use information in a way that encourages escape from the established patterns and insight restructuring into new ones. PO performs a special function that it is impossible to perform adequately in language without PO. Other ways of carrying out this function are cumbersome, weak and ineffective. The more skill and practice one invests in the use of PO the more effective it becomes. It is not language that makes PO necessary but the mechanism of mind.

Blocked by openness 21

I knew the town quite well but I had to ask for instructions as to how to get to this particular restaurant. The instructions were easy to follow as the route was made up of three segments with each of which I was familiar for each of them involved some obvious landmark. The segments had been made familiar by ordinary driving around the town. One day some friends set out for the restaurant from the same place as myself and at the same time. But they got there long before I did. I asked them if they had driven quickly but they denied this. Then I asked them what route they had taken. They explained and it was obvious that they had taken a short cut as shown opposite.

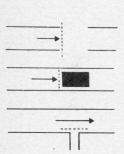

A small side turning had led them directly to the restaurant while I was making an unnecessary detour through the centre of the town. My own route had always seemed satisfactory so I had never looked for a shorter one. Nor had I ever been aware that there was a shorter one. I had driven past the small side turning each time but had never explored it because there had been no reason to explore it. And without exploring it I could never have found out how useful it was. My original instructions had been in terms of large well-known segments of route, cliché segments, because that is the easiest way to give instructions. There had never been any reason to break off along one of these cliché segments. There are three ways in which thinking can be blocked. These three ways are shown diagrammatically.

1 One is blocked by a gap. One cannot proceed further because the road runs out. One needs to find more road or to construct a bridge across the river. This is equivalent to having to look around for more information or having to generate some by experiment.

2 One is blocked by there being something in the way.

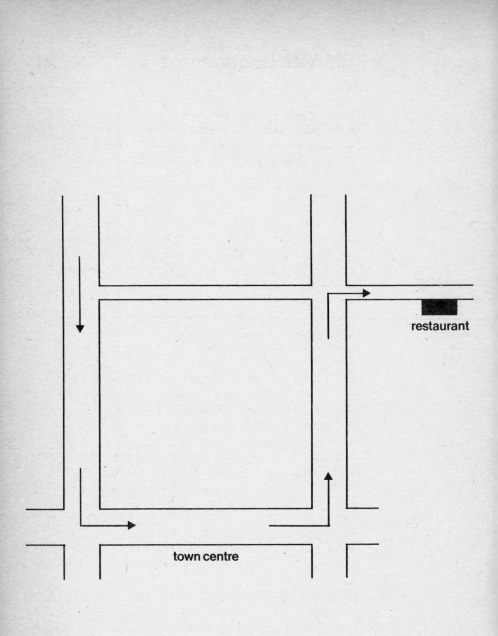

restaurant

town centre

Here there is a definite obstacle which bars progress. In order to go on one has to find a way of removing the obstacle or getting round it. Once this has been done progress is easy because the road is there. One can concentrate one's problem solving efforts on overcoming the block.

3   One is blocked because there is nothing in the way. The road is smooth and clear and so one goes shooting past the important side turning unaware that it is even there. Here a particular way of looking at things leads one straight past a better way of looking at them. Because the first way is adequate one does not even consider that there might be another way – let alone look for it.

This third type of block is what happens when one is blocked by the adequate, blocked by openness. Trying to avoid this sort of block is what lateral thinking is all about. Instead of proceeding with the patterns that have been established on the memory surface of mind one tries to find short cuts to restructure the patterns. Like the route in the restaurant story the established patterns have been constructed out of familiar cliché segments. Even when the patterns are adequate this cannot exclude there being very much more effective patterns.

If things are put together in a certain way to give one pattern then this prevents them being put together in another way to give a different pattern. One way of arranging the three pieces shown opposite excludes the other way. There is an exclusivity about patterns. Nevertheless a satisfactory pattern cannot preclude the possibility of there being a different and better arrangement. The trouble is that the different and better arrangement does not arise from the current pattern but arises instead of it. There is no logical reason to look for a better way of doing something if there is already an

adequate way. Adequate is always good enough. It is
interesting that in our thinking we have developed
methods for dealing with things that are wrong but no
methods for dealing with things that are right. When
something is wrong we explore further. When
something is right our thinking comes to a halt. That is
why we need lateral thinking to break through this
adequacy block and restructure patterns even when
there is no need to do so.

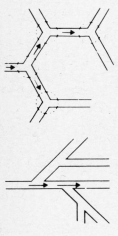

The difficulty with being blocked by openness is that
there is no indication as to where the block has occurred.
It might have occurred anywhere along the apparently
correct pathway. Two types of branching pattern are
shown opposite. In the first type there is a definite
change of direction at each branch point. One has either
to go right or left. This means that one is always *aware*
of the branch points. In the second type of branching
pattern the branches stem off a straight trunk. If you go
along the main pathway you may not even be aware that
there was a side branch or a choice point. One is blocked
by the openness of the main pathway.

If one comes to a dead end in the first type of branching
system one goes back to the branch point and tries the
other branch. This can be done again and again for each
branch point. But in the second type of branching
pattern when one comes to a dead end one cannot just
go back to the preceding branch point because one does
not even know where the branch points are since one
has never had to pause and make a choice at them.

Cliché patterns strung together constitute the trunk of a
straight branching system. As one proceeds smoothly
along them one is not even aware that there are possible
side turnings. So when one comes to a dead end one
does not know where to go.

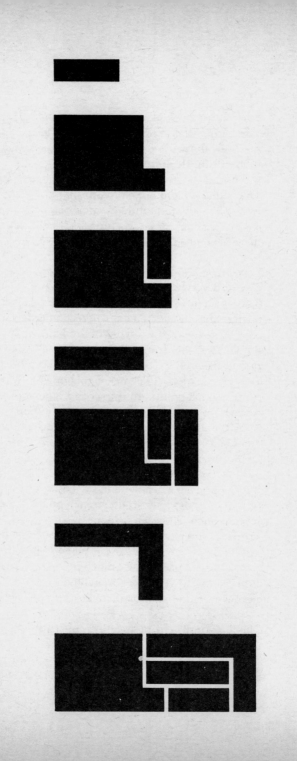

At the top of the opposite page is shown a plastic piece.
Another plastic piece is provided and the task is to
arrange them both together to give a simple shape that
would be easy to describe. The arrangement is obvious
as shown. A further piece is added and once again the
arrangement is obvious. When a fourth piece is added
there is difficulty in fitting them all together. The
original placing of the second piece so that it nestled in
the angle of the first piece is such an obvious pattern
that it becomes a cliché. And as a cliché one wants to
use it not disrupt it. This makes the final solution
difficult since the small piece has to be placed in quite a
different position.

Cliché patterns are satisfactory established patterns
which are very useful and which do a good job. They
can be used in three ways:

1   For communication. It is easier to explain a
situation in terms of cliché patterns than to devise new
patterns.

2   One picks out a cliché pattern more easily than other
patterns from an environment that offers several
alternative patterns.

3   Given only part of a pattern one elaborates this
part to a whole pattern—but a cliché whole.

I was having lunch one day in a university cafeteria
when I noticed sitting at another table a student with
very long hair and a delicate, sensitive face. As I looked
at the student I thought to myself that here was a
person whose sex could not be determined by
appearance. It was several minutes before I suddenly
noticed that the student had a long straggling
moustache! In my mind I had gone at once from the
long hair and delicate face to the assumption that the

student might be a girl and so I had never noticed the moustache. So it is in picking out cliché patterns that one is not even aware that alternative patterns could just as easily have been picked out.

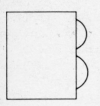

If an ordinary letter is partially hidden under a piece of paper one elaborates the pattern to give the standard letter. Letters are cliché patterns and one only needs a hint in order to be able to elaborate the rest of the letter. It is easy enough to recognize letters in this way because one knows all the possibilities to begin with and also one knows that the pattern must be a letter. But suppose the patterns were not all letters but completely different patterns which were covered up so that the exposed bits did look like letters? One would elaborate the expected cliché pattern and one would be wrong. Or suppose that one did not know the shape of all the letters? The same thing would happen. In real life one is always elaborating patterns as if they could only be standard cliché patterns.

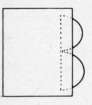

This process of being blocked by openness is very prevalent in thinking. In a way it is the basis of thinking for thinking has to make guesses and assumptions based on past experience. Useful as it is the process has definite disadvantages especially in terms of new ideas and of bringing patterns up to date. This process of being blocked by openness is at the very centre of the need for lateral thinking. Lateral thinking is an attempt to find alternative pathways, an attempt to put things together in a new way, no matter how adequate the old way appears to be.

Practice
The sole intention of this practice session is not to practise any technique but to illustrate the phenomenon of being blocked by openness. This is done by showing how easy it is to be satisfied by what seems to be a satisfactory explanation.

1　Stories, anecdotes, jokes. The students are invited to think of examples of the process of being blocked by adequacy. The examples may be from their own experience or else incidents they have heard about. The teacher can note down these incidents and add them to his own stockpile of material for future occasions. In any case the teacher may already have collected examples of this sort and can use them to illustrate what is wanted.

e.g. I had a guest staying in my house. After the guest had left I found that the reading lamp would not work. I checked the bulb and I checked the fuse but still the lamp would not work. I was just about to dismantle the plug when it occurred to me that the guest might have switched off the lamp by the switch on the lamp base and not the wall switch which is what I usually did. This in fact was what had happened.

2　The students are shown parts of a picture or else a picture with parts obscured by a cardboard sheet. They are asked to decide what the picture is all about. They are encouraged to jump to conclusions before the rest of the picture is revealed.

3　The use of blanks. The students are asked to write a short passage on some theme and then to go over the passage striking out any word which would make the theme obvious. The passage is then rewritten with 'blank' substituted for such words. Alternatively the students can just write the passage and then the teacher strikes out the revealing words and puts blank instead. A third way to do it is to take a passage from a newspaper and magazine and treat it in the same way. It is best to give the students an example of what is wanted before asking them to provide such passages. The blanked out passage is then read out to the rest of the students who are asked firstly to decide what the passage is about and then to try to fill in the individual

blanks. This is done as an individual effort by each
student and at the end the results are compared.

An example of this sort of passage might be:
'He stood by the side of *blank* and every time a *blank*
approached he would raise his arm and *blank*. It was
some time before he eventually got *blank* and even so
that did not take *blank*.'
In this passage the *blank* refers to anything that has
been left out. It is important to point out that this
need not apply to a single word but could be used for a
group of words. Thus the phrase, 'get anywhere' would
be replaced by *blank* just as would be the word 'car'.

The previous section was about being blocked by openness. It dealt with the way adequate established patterns prevented the development of patterns that made better use of the available information. Normally one is only taught to think about things until one gets an adequate answer. One goes on exploring while things are unsatisfactory but as soon as they become satisfactory one stops. And yet there may be an answer or an arrangement of information that is far better than the adequate one. All this is part of the first aspect of lateral thinking. This first aspect is to create an awareness of the limitations of established patterns. Such established patterns can do three things:

1   They can create problems which do not really exist. Such problems are only created by particular divisions, polarizations, conceptualizations.

2   They can act as traps or prisons which prevent a more useful arrangement of information.

3   They can block by adequacy.

This first aspect of lateral thinking is to become aware of the process and the need for it. The second aspect involves developing some skill in the use of lateral thinking.

It is not much use treating lateral thinking as an abstract process. Nor is it much use treating it as something to do with creativity and hence desirable in a general sort of way. Nor is it much use accepting lateral thinking as being of use to some people at some time in some circumstances. Lateral thinking is a necessary part of thinking and it is everybody's business. One needs to go further than awareness and appreciation and to actually practise it. Throughout this book different ways of practising lateral thinking have been suggested.

In each case the idea has been to use a specific technique. In addition to such specific practice sessions one needs some general practice situations. In dealing with the general situations one can use the techniques learned elsewhere or one can develop for oneself deliberate habits of mind and deliberate ways of applying lateral thinking.

It would be possible to become involved in some project in depth. In the course of working through such a project there might be opportunity to use lateral thinking. In fact the opportunity to do so would be very small for in a specialized project treated in depth the emphasis is on the collection of specialized knowledge or its application. This is a matter of vertical thinking. Lateral thinking is most used when knowledge is readily available and the emphasis is on the *best* use of that knowledge. It is far more useful to practise lateral thinking over a large number of small projects than to suppose that it is practised in the pursuit of a large project.

There are three practical situations which encourage the use of lateral thinking.

● Description.
● Problem solving.
● Design.

Description
An object or a situation may be described by someone in a particular way and by someone else in a different way. There can be as many descriptions as there are points of view. Some descriptions may be more useful than others, some descriptions may be more complete than others. But there is no one description which is correct leaving all the others to be wrong. That is why description is an easy way of showing how something can be looked at in different ways. It is also an easy way

of practising the ability to generate alternative ways of looking at something. Furthermore when one learns to generate alternative points of view oneself one is ready to appreciate the validity of other people's points of view.

Description is a way of making visible the way one understands something – the way one explains that thing to oneself. By having to describe something one has to commit oneself temporarily to a particular point of view. This means that one has to generate a definite point of view instead of being satisfied with a vague awareness.

The idea of this exercise is to train people to realize that there is more than one way of looking at a situation and to be able to generate alternative ways for themselves. For this reason the emphasis is not on the accuracy of the description but on the *difference* between descriptions and on the use of novel methods of description.

The raw material which is to be described may be picture material. This could take the form of photographs or ready made pictures or the students could be asked to draw pictures themselves for the others to describe. Simple geometric outline shapes are a good way to start. One can move on from visual material to written material. With written material one is really redescribing something that has already been described. It may be a story, an account from a book or a newspaper article. Real life situations can be identified by name without describing them just as real life objects can be identified and then the full description is left to the students. For instance students could be asked to describe a harvesting machine or the parliamentary system. Acting as in charades could also be the object of description. Obviously there is no limit to what can be described.

The descriptions may be verbal or written or even in picture form. Once they have been obtained the emphasis is on showing the different approaches. Students are encouraged to find still further approaches.

Although one is not interested in finding the best possible description one still needs to bear in mind what is a useful description and what is not. The material to be described is not being used as a stimulus to set off ideas. The task is not to generate ideas which have something to do with the material but to describe that material. The best criterion of adequate description is as follows:
'Suppose you had to describe this scene to someone who could not see it, how would you describe it?'

One is not looking for the complete and pedantic description. A description which only conveys one aspect of the material may be very good if it does so vividly. Descriptions may be partial, complete or general.
For instance in describing a geometric square the following descriptions may be offered:
A figure which has four equal sides.
A figure which has only four angles and all of them are right angles.
A rectangle with all the sides being equal.
If you walk north for two miles then turn sharply east and continue for two miles, then sharply south and continue for two more miles, then sharply west and continue for two more miles, the path of your walk looked at from an aeroplane would be a square.
If you take a rectangle which is twice as long as it is broad and cut it in half straight down the middle you would have two squares.
If you put together two right-angled isosceles triangles, base to base, you would have a square.

Some of the above descriptions are obviously very incomplete. Others are very roundabout.

Description is certainly the easiest setting in which to practise lateral thinking because there is always some result.

Problem solving

Like description problem solving is a format that has been used in the suggested practice sessions throughout the book. A problem is not just an artificially arranged difficulty that is only to be found in textbooks. A problem is simply the difference between what one has and what one wants. Any question poses a problem. Generating and solving problems is the basis of forward thinking and progress. If description is a matter of looking back to see what one has then problem solving is a matter of looking forward to see what one can get.

In any problem there is a desired end point – something one wants to bring about. What one wants to bring about may take a variety of forms:
1   To resolve some difficulty (traffic congestion problem).
2   To bring about something new (design an apple picking machine).
3   To do away with something unsatisfactory (road accidents, starvation).
All these are but different aspects of the same process which is to bring about a change in the state of affairs. For instance the traffic congestion problem could be phrased in three ways:
1   To resolve the difficulty of traffic congestion.
2   To design a road system which would have free traffic flow.
3   To get rid of the frustration and delay of traffic congestion.

Problems may be open ended or closed. Most of the
problems used in this book are open ended problems.
This is because it would be impossible to have the time
or the facilities for trying out solutions to a variety of
real life problems. With open ended problems one can
only offer suggestions as to how the problem might be
solved. Since these suggestions cannot actually be tried
out to see if they work they have to be judged in some
other way. Judgment is based on what one thinks would
happen if the solution was actually tried out. It may be
the teacher who makes the judgment or the other
students. The emphasis, however, is not on judging the
suggested solutions but on generating different
approaches. Where possible one acknowledges a
suggestion and even elaborates it rather than reject it.
The only time one has to enforce judgment is when the
suggestions wander so far from the problem that one is
no longer trying to solve it at all. Though a problem
may in fact be solved by information generated in
another context the purpose of this type of problem-
solving practice is to try to solve the given problem.

With closed problems there is a definite answer. The
solution either works or it does not. There may be only
one solution but more often there are alternative
solutions. Some of these solutions may be better than
others but for this purpose it is enough that the solution
works. It is better to find a variety of solutions than to
only find the best one. Closed problems have to be
fairly simple because they have to be capable of being
solved in a simple setting. Alternatively one has to have
a notational system like mathematics which permits one
to make one's own model of the real world. It is better
however to keep away from purely mathematical
problems since these require knowledge of technique.
There are various verbal problems which have verbal
solutions. Some of them involve the simplest of
mathematics but the solution really depends on the way

the problem is looked at. (e.g. There was a line of ducks walking along and there were two ducks in front of a duck and two ducks behind a duck. How many ducks were there? The answer is three ducks.) One can build up a stock of such problems by noting them down whenever one comes across them. It is very important that none of the problems depends on verbal tricks for the students must not be given the impression that the teacher is out to *trick* them by means of puns and so on.

A useful type of problem is the artificial mechanical problem of the closed type. Such problems deal with actual objects, for instance how to get a long ladder through a short room. It is possible to generate such problems deliberately by taking a simple straightforward activity and then making a problem of it by severely limiting the starting position. For instance the problem might be: 'How would you empty a glass of water if you are not allowed to lift it off the table?' Another such problem might be: 'How could you carry three pints of water in a newspaper?' When using this type of problem one must be extremely careful in defining the starting position. One cannot go back afterwards and say that something was assumed or taken for granted. For instance if you ask students to cut a postcard into a certain shape then you cannot say: 'But I did not say you could fold the card,' or 'It was assumed you could not fold the card otherwise it would be too easy.' This point is important because if you tell students to make assumptions and presume boundaries in their problem solving then you are going directly against the purpose of lateral thinking which challenges the limiting effect of such assumptions.

Many of these artificial closed problems may seem rather trivial. But this does not matter for the processes used in solving such problems can be isolated and transferred to other problems. The idea is to develop a repertoire of problem solving processes.

There is a third type of problem which can be used in the classroom situation but it involves the teacher doing some homework. The idea is to put forward problems that have already been solved but to withhold the solution. The teacher has to imagine how the problem might have been stated before the solution was found. The situations must of course be ones with which the students are not familiar. For instance students might be asked: 'How would you make plastic buckets or plastic tubing?' The teacher who would know about moulds, vacuum forming, extrusion etc would encourage suggestions and give the answer at the end. It is sometimes as well to ask if anyone already knows the answer because if so he can be told to keep quiet or to explain the answer himself at the end. If the students each write out their own suggestions there is no danger of the problem being spoiled by someone who knows the answer. This sort of problem can be generated by using one's imagination, by reading magazines (science, technology etc) or by wandering around exhibitions. There is no harm in reinventing things that have already been invented. It is very good practice.

Design

Design is really a special case of problem solving. One wants to bring about a desired state of affairs. Occasionally one wants to remedy some fault but more usually one wants to bring about something new. For that reason design is more open ended than problem solving. It requires more creativity. It is not so much a matter of linking up a clearly defined objective with a clearly defined starting position (as in problem solving) but more a matter of starting out from a general position in the direction of a general objective.

A design does not have to be a drawing but for the practice of lateral thinking it is much more useful if the design always takes the form of a drawing. It does not matter how good the drawing is so long as there is an attempt to give a visual description of what is meant.

Explanatory notes may be added to the drawing but they must be brief. The advantage of a drawing is that there is far more commitment than with a verbal explanation. Words can be very general but a line has to be put in a definite place For instance in the design of a potato peeling machine it would be easy to say, 'The potatoes go in there and then they get washed.' But when this is described visually one can get the effect shown opposite. The designer wanted to use a bucket of water to wash the potatoes and the best way to fit the bucket into his machine was by turning it on its side – so the water level had to be turned on its side as well. This beautiful cliché use of the bucket of water would never have been apparent in a purely verbal description.

● Comparison
The first purpose of the design exercise is to show that there are alternative ways of carrying out some function. A single designer will only be able to see one or perhaps a few alternative ways of doing something. But with a large number of designers there will be a large number of alternative approaches. Thus by simply exposing any single designer to the efforts of the others one shows how it is possible to look at things in different ways. The object of the design session is not to teach design but to teach lateral thinking – to teach the ability to generate alternative ways of looking at something.

In practice some general design theme is given to the class (apple picking machine, cart to go over rough ground, potato peeling machine, cup that does not spill, redesigning the human body, redesigning a sausage, redesigning an umbrella, a machine to cut hair etc). The students are asked to come up with designs for the particular design task set. To make comparison easier it is best to only set *one design project* rather than let the students select their own from a list.

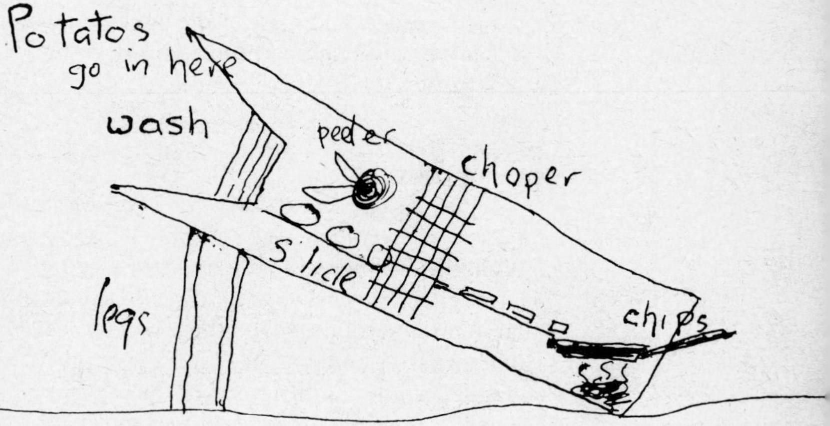

Potatos
go in here

wash

peder

choper

slide

legs

chips

Potato mashin

The individual designs are then collected and
compared.

The comparisons may refer to the whole design (e.g.
picking the apples off a tree as compared to shaking the
tree) or to some particular function (e.g. grabbing the
apples with a mechanical hand as compared to sucking
them through a hole).

● Cliché units
In examining the submitted designs one very quickly
becomes aware of cliché units. Cliché units are
standard ways of doing something that are borrowed
entire from another setting. For instance a bucket and
water to wash potatoes in is a cliché unit. The second
purpose of the design exercise is to point out these
standard ways of doing things and to show how they
may not be the best way.

In pointing out the cliché units one does not judge
them. Certainly one does not condemn them for being
cliché units. In the design process one has to go
through cliché units before moving on to something
more appropriate. One merely points out the cliché
unit and encourages the designer to go further.

The entire design may be a cliché unit. Thus when
children were asked to design a cart to go over rough
ground one boy drew a warlike tank complete with
cannon, machine guns and rocket missiles. Such
entire cliché units are borrowed directly from films,
television, comics, encyclopedias etc.

More often the cliché unit is only part of the design. In
the apple picking machine project one student drew a
large robot man picking apples off a tree. From the top
of the robots head a wire went to a control switch in the
hand of a normal man standing just behind. The

This macinery go over rough ground.
This is a difirnet kink of tank it has a misilils
oner bomb two cons. This tank gos over rough ground
There is three men wokm woking the tank
It will push a rock out the wae

large robot was complete down to eyelashes. Another
picture showed a boxlike structure with a plain disc for
a head. This structure stood on two legs and it was
equipped with two simple picking arms each of which
had five fingers. Another design had done away with the
legs and converted the disc like head into a dial with a
pointer showing 'fast . . . faster . . . stop'; but the two
arms with five fingers were retained. A further design
did away with the head but kept the arms. Finally a
very sophisticated design showed a small mobile
wheeled car with a long arm that stretched out to the
apples. At the end of the arm was a complete hand with
five fingers. One might have supposed this was just a
neat way of indicating a picking function but there
was a black hole in the middle of the hand and an
explanatory note, 'apples are sucked through this hole'.
In this series cliché units ranged from the complete
duplicate man to the hand with five useless fingers.

As suggested above one may have to pass through
cliché units in the course of the design process. The
cliché units may be handled in the following ways
(among others).

1   Trimming and splitting:
A complete cliché unit is taken and then the inessentials
are trimmed away much as one might trim a rosebush.
For instance in a sophisticated design for a potato
peeling machine one designer wanted to go further and
fry the potatoes to make chips. So he included a frying
pan *complete* with handle. Since the potatoes were
mechanically transported into and out of the pan the
handle was obviously superfluous.

Through repeated trimming one gradually narrows
down the cliché unit to that part which is really
necessary. (This is the whole purpose of that branch of
engineering known as Value Engineering.) Trimming

may be a gradual process with a small amount removed
each time or it may involve large slashes. For instance
from the cliché unit of a tank one may slash off all the
warlike function and keep only the caterpillar track.
Where the jump is very large it may be more a matter of
splitting a cliché unit than trimming it. Trimming and
splitting are concept breaking procedures and being
able to use them is a process of lateral thinking – the
escape from rigid patterns.

2   Abstraction and extraction:
In a way this is just a form of splitting. To extract the
critical part of a cliché unit is the same as splitting off
everything else. In practice however the two processes
are different. One may either recognize the essential
part and remove it (extraction) or one may deal with the
cliché unit trimming off bit after bit until one comes to
the essential part.

What is extracted may actually be part of the cliché unit.
On the other hand it may be something less tangible,
something that depends on looking at the cliché unit in
a particular way. For instance one may *abstract* the
concept of function. Though the concept is derived
from the cliché unit it is not a physical part of it but a
particular description. Nevertheless it might not have
arisen without the cliché unit. Thus in the apple
picking machine 'picking' is an abstracted function that
arises directly from the cliché unit of the human hand.

3   Combining
Here one takes cliché units from several different
sources and puts them together to give a new unit
which does not occur anywhere. This process of
combination may be by simple addition of function
(caterpillar tracks, telescopic arm, hand to pick apples)
or there may be some multiplication of function (e.g.
for a redesign of the human body: noses on the legs so

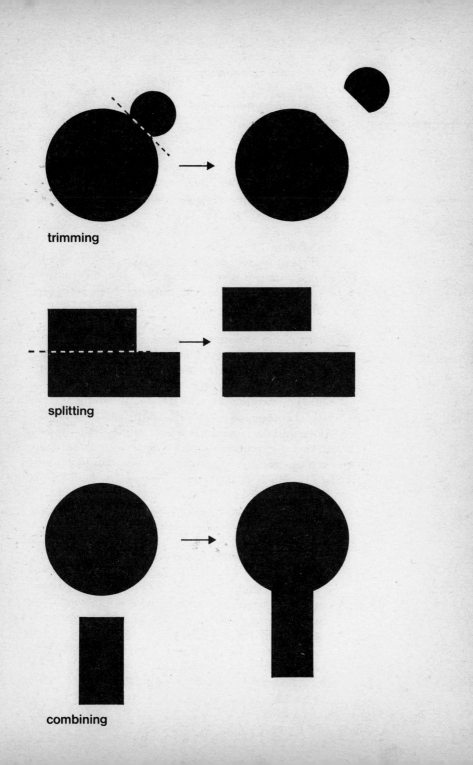

trimming

splitting

combining

they would be nearer the ground and be more useful for tracking).

These different ways of handling cliché units cover the basic processes of selection and combining which are of course the basis of any information processing system. The processes are shown diagrammatically on the previous page.

● Function
As distinct from objects function is the description of what is happening, what is going on. It is easy to think of particular objects or arrangements of objects as clichés but functions can be clichés as well.

In any design situation there is a hierarchy of ways of looking at the function. One could proceed from the most general description down to the most specific. For instance in the apple picking machine situation one could have a hierarchy which went something like this: getting apples to where you want them, separating the apples and the tree, removing the apples from the tree, picking the apples. Normally one does not go through such a hierarchy but uses a specific description of function such as 'picking the apples'. The more specific the description the more one is trapped by it. For instance the use of 'picking' would exclude the possibility of shaking the apples off the tree.

In order to escape the trap of a too specific idea of function one tries to go backwards up the hierarchy of function, from the specific to the more general. Thus one would say, 'not picking apples but removing apples, not removing apples but separating apples from the tree.' Another way of escaping from the too specific idea of function is to change it around in a true lateral manner. Thus instead of 'picking the apples from the tree' one would think of 'removing the tree from around the apples'.

When asked to design a cup that would not spill a group
of children showed a variety of functional approaches.
The first approach was to design a cup that could not be
knocked over. Three possible ways of doing this were
suggested: long hands that descended from the ceiling
to immobilize the cup; 'sticky material' on the table to
attach the cup; a pyramid shaped cup. The second
approach was to have a cup that would not spill even
when it was knocked over. This was done either by
having a special cover to the cup (the cover being
flipped open by a catch when one wanted to drink) or by
shaping the cup so that the liquid always stayed at the
bottom no matter in what position the cup was (rather
like unspillable inkwells).

The trouble with function is that once one has decided
the particular function then one's design ideas are very
much fixed. So one wants to pay attention to
generating alternative *functions* and not just ways of
carrying out a particular function.

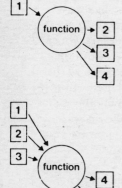

The abstraction of a function is a very useful way of
getting ideas moving in the design process. If one is
stuck with a particular way of doing something (a hand
to pick apples) then one can not get much further. But
if one abstracts the function from this particular
situation then one can find other ways of carrying out
that function. This process is shown in the diagram
opposite. The design results obtained from the
students can be compared by showing which designs
are but different ways of carrying out the same function.
On the other hand one can also show how a different
concept of function leads to a completely different
approach.

In dealing with function one wants to show two things:
1   How the abstraction of a function can lead to
different ways of carrying out this function.

2   How one may need to change a particular idea of function in order to generate new approaches.

In practice one might say: 'That is one way of carrying out this picking function – can you think of any others?' But one might also say: 'Those are different ways of carrying out this picking function but is that the only way of looking at it. Suppose we leave aside the idea of picking and just think of removing the apples from the trees.'

● Design objectives
In a design problem there is very rarely only a single objective. Usually there is a main objective and many subsidiary objectives which may not be apparent. For instance in the design of an apple picking machine the main objective may be to reach and pick the apples but in achieving this objective one may make it impossible to achieve the other objectives as well. Shaking the trees to remove the apples would satisfy the main objective but it would damage the apples. Having a huge machine to do the job might satisfy both the above objectives but might be so uneconomical that it would still be cheaper to do it by hand. Thus three objectives have become apparent: picking the apples, obtaining undamaged apples, a machine that is more economical to use than hand labour. There are other objectives. For instance the machine might have to work at a given speed or it might have to be of such a size that it could pass easily between the trees in a standard orchard. All these objectives might be specified in a description of a desired machine or else they might only become apparent when the design was being examined.

Some designers try to keep all the objectives in mind all the time. They would only move forward very slowly and they would immediately reject an idea that failed to satisfy one of the objectives. Other designers would move quickly ahead in an attempt to satisfy the main objective. Having found some sort of solution they would then look around and see how well the other objectives were satisfied. This second method is probably more generative but it does depend on a thorough assessment at the end otherwise the effect may be disastrous if one important objective is overlooked. It is better to have this assessment at the end rather than at each stage for an assessment at each stage would prevent the consideration of ideas which were inadequate in themselves but served as stepping stones to much better ideas.

● Design and lateral thinking
This section is not meant to be a treatise on design but an indication that the design process involves much lateral thinking and provides an excellent setting in which to practise lateral thinking. In the design process one is always trying to restructure concepts; one observes cliché units and tries to get rid of them; one is continually having to generate fresh approaches.

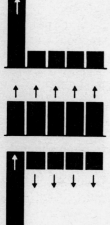

Many of the examples used in this section were obtained from the design efforts of children aged seven to ten years old. Such children are relatively unsophisticated and the design process is a caricature of the design process which would be used by older people. The advantages of such examples is that the design process and its faults are made much more clear. The faults arise from the way the mind handles information and not from any peculiarity of youth. In a less obvious form the same faults occur at all age levels.

The first purpose of the design setting is to get students to generate alternatives. The second purpose is to get them to look beyond the adequate in order to produce something better. The third purpose is to free them

from domination by cliché patterns. These three
purposes paraphrase the purpose of lateral thinking.

Practice

The students are set a specific design task. Each student
tackles the same task. Every design is a drawing. Brief
notes may appear on the drawing to indicate how
something works. In addition there may be a fuller
explanation but this fuller explanation should only refer
to what is already in the drawing – it is not to be a
substitute for the drawing. Half an hour is enough
time to allow for each design project since one is not so
interested in the excellence of the design but in the
process itself.

When the design task is set some of the students may
ask for additional information. For instance if the task
was to design a vehicle to go over rough ground then
someone could well ask how rough the ground was to
be. Though such questions are perfectly legitimate and
in a real design situation one would specify the
objective very closely it is better to specify nothing. This
means that each student is allowed to assume his own
specifications. This gives a much wider variety of
response. In discussing the results one can comment on
the way the designs fulfil other objectives as well as the
main one but one must not condemn a design for not
fulfilling a condition which was never given.

The collected results may be discussed there and then
or they may be examined and discussed at a subsequent
session. Wherever possible it might be an advantage to
display the results in some way before discussing them.

As suggested before the discussion is centred on
comparisons of the different ways of doing things and
the picking out of cliché units. It is best to avoid making
comparisons as to which is the best design for fear of

restricting imagination. If one does want to pick out a
design as being very good one can do so by commenting
on something specific for instance the originality or
economy of it rather than giving a blanket approval such
as 'good'. Otherwise one uses such comments as
'interesting', 'unusual', 'very different' etc. Above all
one wants to refrain from condemning any particular
design. Such condemnation can only be restrictive. If
one wants to encourage some particular feature one can
do so by praising it where it is present rather than
condemning its absence. For this reason it is best not to
allow students to pass open judgment on the design
efforts of others (i.e. not to call for such judgments in
the class situation).

Suggestions for design projects have been given in the
course of this section. In general the design project
may ask for a design to do something that is not
done at the moment (e.g. a machine to cut hair), or to do
something in a better way (e.g. redesign a comb). The
projects may be simple or more complicated. On the
whole simple mechanical designs are more useful than
abstract ideas. Students may be asked to redesign any
everyday object whatsoever, for instance: telephone
receiver, pencil, bicycle, stove, shoes, desks. Further
suggestions are given in the previous section on design.

● Will it work?
One does not want to restrict designs to the sane,
workable ones by carefully analysing each one and
rejecting those which would not work. Nevertheless one
does want the students to aim for a workable design and
not produce a fantasy for the sake of fantasy. The level
of mechanical knowledge which one could expect of the
students obviously varies with their age but in any case
one is not testing this. It is sufficient if every now and
again the teacher picks out a design which would

obviously not work and gets the class as a whole to accept that it would not work but can still lead to useful ideas. The judgment is not as to whether the design is workable but as to whether the designer was genuinely trying to make a workable design (even if everyone else can see that it would not work). If there is any doubt it is better to say nothing and simply ignore the design.

Summary

The emphasis in education has always been on logical sequential thinking which is by tradition the only proper use of information. Creativity is vaguely encouraged as some mysterious talent. This book has been about lateral thinking. Lateral thinking is not a substitute for the traditional logical thinking but a necessary complement. Logical thinking is quite incomplete without lateral thinking.

Lateral thinking makes quite a different use of information from logical (vertical) thinking. For instance the need to be right at every step is absolutely essential to logical thinking but quite unnecessary in lateral thinking. It may sometimes be necessary to be wrong in order to dislocate a pattern sufficiently for it to re-form in a new way. With logical thinking one makes immediate judgments, with lateral thinking one may delay judgments in order to allow information to interact and generate new ideas.

The twin aspects of lateral thinking are first the provocative use of information and second the challenge to accepted concepts. Underlying both these aspects is the main purpose of lateral thinking which provides a means to restructure patterns. This restructuring of patterns is necessary to make better use of information that is already available. It is an insight restructuring.

The mind is a pattern making system. The mind creates patterns out of the environment and then recognizes and uses such patterns. This is the basis of its effectiveness. Because the sequence of arrival of information determines how it is to be arranged into a pattern such patterns are always less than the best possible arrangement of information. In order to bring such patterns up to date and so make better use of the contained information one needs a mechanism for insight restructuring. This can never be provided by logical thinking which works to relate accepted concepts not to restructure them. Lateral thinking is demanded by the behaviour of this type of information processing system in order to bring about insight restructuring. The provocative function of lateral thinking and the challenging function are both directed towards this end. In both cases information is used in a manner that goes beyond reason for lateral thinking works outside of reason. Yet the need for lateral thinking is based quite logically on the deficiencies of a self-maximizing memory system which is the type of system that makes the mind capable of humour.

Lateral thinking works at an earlier stage than vertical thinking. Lateral thinking is used to restructure the perceptual pattern which is the way a situation is looked at. Vertical thinking then accepts that perceptual pattern and develops it. Lateral thinking is generative, vertical thinking is selective. Effectiveness is the aim of both.

In ordinary traditional thinking we have developed no methods for going beyond the adequate. As soon as something is satisfactory our thinking must stop. And yet there may be many better arrangements of information beyond the merely adequate. Once one has reached an adequate answer then it is difficult to proceed by logical thinking because the rejection mechanism which is the basis of logical thinking can no longer function well. With lateral thinking one can easily proceed beyond the adequate by insight restructuring.

Lateral thinking is especially useful in problem solving and in the generation of new ideas. But it is not confined to these situations for it is an essential part of all thinking. Without a method for changing concepts and bringing them up to date one is liable to be trapped by concepts which are more harmful than useful. Moreover rigid concept patterns can actually create a great number of problems. Such problems are particularly fierce since they cannot be altered by available evidence but only by insight restructuring.

The need to change ideas is becoming more and more obvious as technology speeds up the rate of communication and progress. We have never developed very satisfactory methods for changing ideas but have always relied on conflict. Lateral thinking is directed towards bringing about changes in ideas through insight restructuring.

Lateral thinking is directly concerned with insight and with creativity. But whereas both these processes are usually only recognized after they have happened lateral thinking is a deliberate way of using information in order to bring them about. In practice lateral thinking and vertical thinking are so complementary that they are mixed together. Nevertheless it is best to treat them as distinct in order to understand the basic nature of lateral thinking and acquire skill in its use. This also prevents confusion because the principles governing the use of information in lateral thinking are quite different from the ones used in vertical thinking.

It is difficult to acquire any sort of skill in lateral thinking simply by reading about it. In order to develop such skill one must practise and go on practising and that is why there has been such emphasis in this book on practice sessions. Nor are exhortation and goodwill enough. There are specific techniques for the application of lateral thinking. The purpose of such techniques is twofold. They can be used for their own sakes but more importantly they can be used to develop the lateral habit of mind.

In order to use lateral thinking effectively one needs a practical language tool. Such a tool is necessary to allow one to use information in the special way required by lateral thinking and also to indicate to others what is being done. This tool is PO. PO is an insight tool. PO is the laxative of language. It acts to relax the rigidity of the tight patterns so easily formed by mind and to provoke new patterns.

Lateral thinking is not concerned with generating doubt for the sake of doubt or chaos for the sake of chaos. Lateral thinking acknowledges the extreme usefulness of order and of pattern. But it emphasizes the need for changing these to bring them up to date and make them even more useful. Lateral thinking particularly emphasizes the dangers of rigid patterns which the mind is so apt to construct because of the way it handles information.